THE PHYSICS OF EVERYDAY PHENOMENA

Readings from
**SCIENTIFIC
AMERICAN**

THE PHYSICS
OF EVERYDAY
PHENOMENA

With an Introduction by
Jearl Walker
Cleveland State University

W. H. Freeman and Company
San Francisco

Most of the SCIENTIFIC AMERICAN articles in *The Physics of Everyday Phenomena* are available as separate Offprints. For a complete list of articles now available as Offprints, write to W. H. Freeman and Company, 660 Market Street, San Francisco, California 94104.

Library of Congress Cataloging in Publication Data

Main entry under title:

The Physics of everyday phenomena.

Bibliography: p.
Includes index.
1. Physics—Addresses, essays, lectures. 2. Meteorology—Addresses, essays, lectures. 3. Oceanography—Addresses, essays, lectures. I. Walker, Jearl, 1945-
II. Scientific American.
QC71.P47 1979   551.5   79–9287
ISBN 0–7167–1125–7
ISBN 0–7167–1126–5 pbk.

Printed in the United States of America

9 8 7 6 5 4 3 2 1

Cover photograph: Plate 35 from *Field Guide to Snow Crystals* by Edward R. LaChapelle (Seattle: University of Washington Press, 1969).

CONTENTS

Note on cross-references to SCIENTIFIC AMERICAN *articles:* Articles included in this book are referred to by title and page number; articles not included in this book but available as Offprints are referred to by title and offprint number; articles not included in this book and not available as Offprints are referred to by title and date of publication.

THE PHYSICS OF EVERYDAY PHENOMENA

INTRODUCTION

Nowadays the headlines in physics are usually grabbed by explorations into deep space or investigations of mysterious aspects of fundamental particles. The work always seems to require elaborate instruments or huge machines, such as radio telescopes or particle accelerators.

But physics has not always been like that, and still is not for most working physicists. The intriguing puzzles scientists usually solve are those lying about in plain sight in the everyday world, often things so ordinary that they are overlooked except by keen observers. The art of such clear inspection is largely neglected in the training of new scientists. They are more likely to pore over scribbled calculations or ordered numbers on a computer printout than to notice the real world going on about them.

One of the most acute observers in the history of science was Marcel Gilles Minnaert, who early in this century wrote the definitive account of optics in the natural landscape: *Light and Colour in the Open Air*. When someone mentioned the red or blue sky to Minnaert, not only would he describe the associated Rayleigh scattering of light by atmospheric molecules, but he would also point to the sky, calling attention to the subtle variations in color. The sky is not just red or blue. There are times when the sky should be one or the other, but in fact is neither. There are other times when the sky should be neither, but in fact is deep blue. For Minnaert the world was an ongoing laboratory that merited constant attention. And for him the first step in understanding how the world works was careful observation.

This collection of articles from *Scientific American* deals with ordinary things. Each article treats a natural phenomenon that is so common as to be ignored by most of us in the complexity and confusion of daily life. Usually we think that the workings of these phenomena were figured out long ago and that the physics of them has been relegated to the back shelf of some dusty museum devoted to old (and useless) physics. I find these articles fascinating because they show that just the opposite is true. The mysteries of these phenomena may be a shade better understood than they were fifty years ago, but they are still intriguing, still demanding the attention of an acute observer.

In several cases, the analysis of the physics involved in the phenomena is more art than science. For example, if someone asked you to describe the shape of a falling raindrop, what would you answer? You have seen drawings and cartoons of raindrops for most of your life. They are, of course, tear-shaped to reduce aerodynamic drag as they fall, right? Wrong! Well, if not tear-shaped, then surely they must be spherical because surface tension works to reduce the surface area to a minimum, given the fixed volume of the drop. So raindrops are spherical, right? Wrong, again. The fact is that raindrops are not spherical either, despite the logical basis of our intuitive guess. As pointed out in James E. McDonald's article, the actual shape of raindrops was not thoroughly investigated until the middle of this century, largely because the intuitive ideas were so firmly entrenched that nobody thought to examine their shape scientifically.

The shapes of snow crystals (B. J. Mason's article) have fascinated scientists and laymen alike for centuries—long after the basic hexagonal shape of the crystal was understood. The problem of snow crystals would not be so intriguing if there were only a single shape. The mystery lies in the variety of shapes they can assume: hexagonal plates; needles; hollow, prismatic columns; branched, star-shaped crystals. What happens to falling water that determines its final shape? As water molecules from the air accumulate on the crystal, how does the shape evolve? In particular, why are snow crystals symmetrical? How do the molecules clinging to one side of the crystal "know" what shape the molecules on the other side are building?

As researchers probed these questions, they uncovered other mysteries. They found that water droplets in clouds were cooled well below the normal freezing point of water; yet they remained liquid. One of the factors influencing this supercooling of water turned out to be that there were only a few particles onto which the water could freeze. As soon as the role of the nucleating particles was discovered, scientists attempted to seed clouds to cause precipitation over arid land. The questions of which particles best increase precipitation and how water is finally released from clouds in such seeding are still being explored by scientists. And their solutions are becoming increasingly important as the world runs short of food and depends more critically on the weather.

The action of running water can create the graceful curving of rivers (the article by Luna B. Leopold and W. B. Langbein). This effect is more easily observed now that air travel is commonplace. Why should rivers, especially older rivers, weave back and forth? The question has teased a great many people for at least the past two hundred years. Einstein wrote an article on the subject, relating river meandering to the circulation system in a briskly stirred cup of tea. The forces that eat away at the outer edge of a curving river bank are the same forces that carry tea leaves along the bottom of a teacup from the outside edge to the center. From the meandering rivers to the flow in a common teacup, the astute observer of nature can discern the simple, common mechanism that underlies apparent diversity.

Understanding the creation and propagation of water waves (Willard Bascom's article) is not easy; laymen have sought to comprehend their nature for centuries, and scientists have devoted a lifetime to their study. Even the creation of tiny water waves from breezes playing over the surface of a pond was not understood until fairly recently. More mysterious still have been the large, violent, and deadly waves that sweep across the oceans virtually unnoticed until they suddenly demolish a coastal town. Partly to save lives and property and partly to understand the oceans in general, scientists have turned more and more attention to ocean waves. There have been times, during wars, that precise knowledge of coastal waves and shorelines was critical. But even during peacetime such knowledge is important because our future depends in part on the fate of the oceans. Our dependence ranges from the use of food and material resources available from the oceans to the subtle yet crucial role the oceans play in maintaining our atmospheric environment. Man is by nature a tinkerer, but inappropriate tinkering with either the atmosphere or the oceans would be catastrophic.

I noted earlier that people rarely raise questions about ordinary phenomena. An obvious question to be asked about the ocean is: Why is it salt? Does the salt come from the rivers draining into the ocean? Is the salt somehow produced in the ocean itself? Is there a loss of salt to the atmosphere? These questions (discussed in Ferren MacIntyre's article) are springboards to the understanding of the complicated relationship among the oceans, the atmosphere, the rivers, precipitation, and the movement of tectonic plates.

A dramatic example of human tinkering with the world without fully understanding the possible results is the smog that hangs over large industrial cities (Joel N. Myers' article). Residents of Los Angeles suffer through a persistent smog that is aggravated by automobile emissions and other sources of pollution. It is thought that smog was present before the advent of industry, so industrial pollution is only partly to blame. The open burning of coal in London has led to more disastrous results: on one occasion the fog it created killed about 4,000 people. Evidently it was not understood that the water drops forming the fog nucleated on the airborne particles resulting from the burning of coal.

The formation of such fog affects the heat balance of the earth's atmosphere. Our atmosphere is in a precarious balance. Any large-scale alteration of its components could change our environment drastically. What the world needs—and quickly—are keen observers who can detect possible trouble before disaster results. In recent years scientists discovered that chlorofluorocarbons released from such apparently harmless things as spray deodorants could significantly reduce the ozone shield protecting us from the sun's ultraviolet rays. In this instance a serious hazard was detected, and corrective action taken, in time to avert a potential disaster.

Even casual observers have been intrigued by the northern light, the auroras (Syun-Ichi Akasofu's article). These shimmering ribbons, curtains, folds, bundles, and patches, in a rainbow of colors, have surely fascinated people since ancient times. The displays were probably originally thought to be associated with the unpredictable behavior of the gods; in more recent times, with the unpredictable behavior of the earth's atmosphere. By the middle of this century, physicists began to realize that the auroras were linked not so much to the earth's atmosphere but to the sun. The full details of the link will still take some time to work out, for it involves the complicated structure of the atmosphere at high altitudes, the earth's magnetic field, and the particle streams spewed from the sun. There is some evidence that a similar link exists between the sun and our daily weather. For example, some physicists have claimed that the frequency of thunderstorms is linked to the density of sunspots.

One article in this collection deals with the aspect of thunderstorms that I have always found the scariest: thunder (Arthur A. Few's article). A quick flash of light is thrown across my bedroom wall, and then, a short time later, a hair-curling boom of thunder. You probably already know why the thunder arrives later than the light; but did you ever listen carefully to the thunder? Sometimes it is just a clap; at other times it is a prolonged roll—as if the gods were rolling their chariots across the stormy sky. Why the variation? Why is it that sometimes lightning produces no thunder so far as you can tell? In general, you can't hear the thunder of a lightning stroke that is more than fifteen miles away. These questions and others have been answered in detail only in the past twenty years.

This collection of articles deals with only a small outcrop on the mountain of observations one can make in the real world. Phenomena like thunder, that go largely unstudied, are gateways to understanding how the world works, but only if one becomes observant, attentive, and curious enough to investigate them.

The Growth of Snow Crystals

by B. J. Mason
January 1961

Much of the world's precipitation is triggered by natural dusts that act as nuclei in causing water droplets in clouds to freeze. Some artificial nuclei work more effectively than natural ones.

The remarkable beauty of snow crystals, revealed in the classic elegance of their simple geometrical shapes and the delicate tracery of their more intricate forms, has long been recognized and recorded by the scientist, the artist and the industrial designer. It is only in recent years that a serious scientific study has been made of their structure, germination and growth. These studies have been largely motivated by the increasing interest in the physics of clouds and the formation of rain, and in the possibility of modifying these processes artificially. It appears that over large portions of the earth raindrops first begin their lives as snow crystals; then they melt before they reach the ground.

My colleagues and I at the Imperial College of Science and Technology in London have spent a number of years studying the birth and growth of snow crystals in the laboratory, hoping to learn something about the way the crystals develop in clouds. Except for the very cold, high-altitude cirrus types, which are thin veils of small ice crystals, clouds consist mainly of water droplets so tiny and so dispersed that they stay suspended in the air like smoke particles. For years meteorologists puzzled over this stability of clouds, and were hard pressed to explain how the tiny water droplets ever grow large enough to fall as rain. It was equally puzzling that the water droplets often refuse to freeze even though the cloud may be many degrees below the nominal freezing point of water: zero degrees centigrade, or 32 degrees Fahrenheit. Even on a hot summer day the temperature of the air above 15,000 feet is usually below freezing.

During the 1930's Tor Bergeron of Sweden and Walter Findeisen of Germany provided a theory of cloud be-

havior that seems to account satisfactorily for much of the world's precipitation. They proposed that clouds remain stable until a small percentage of the cloud droplets finally freeze, spontaneously or otherwise. When water molecules are locked into place in an ice crystal, they evaporate much less readily than they do from a drop of water. Thus if a cloud contains both water droplets and ice crystals, the water molecules that diffuse from the vapor state onto the ice crystals tend to be bound fast, and those that condense on the water droplets are relatively free to evaporate again. As a result the crystals grow more rapidly than the droplets; finally, as the air is denuded of moisture by the ice crystals, the water droplets evaporate and disappear. The ice crystals meanwhile grow

large enough to fall toward the earth. After growing for about an hour in a deep layer of cloud, a snow crystal will reach the size of a drop in a drizzle, or perhaps the size of a small raindrop. Such crystals fall at the rate of about one foot per second. Several of them may become joined together, as they settle through the air in a fluttering or tumbling motion, to form a snowflake which, in falling into the warmer regions of the cloud, may melt and reach the ground as a raindrop.

Bergeron and Findeisen originally believed that virtually all the world's precipitation—snow or rain—originated with this ice-crystal mechanism, but it is now known that, especially in the tropics, rain sometimes falls from clouds so warm that ice could never have formed

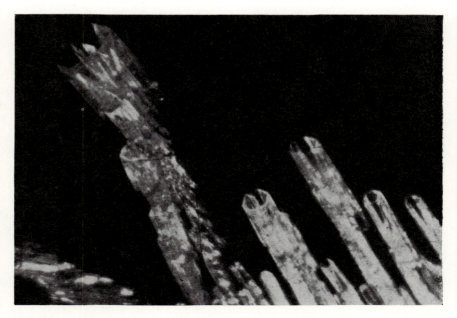

THREE BASIC FORMS OF SNOW CRYSTALS provide the basis for an infinite variety of shapes. Hollow prismatic columns (*left*) populate cirrus clouds, which are usually colder

in them. Findeisen proposed as early as 1938 that for ice crystals to appear in supercooled clouds a nucleating or seeding agent might be required. He suggested that the agent might be dust particles of the proper configuration to start the nucleation of snow crystals, but he was never able to demonstrate its existence.

The subsequent history of cloud-seeding is well known. In 1946 Vincent J. Schaefer, then working at the General Electric Research Laboratory, discovered that ice crystals could be nucleated in a supercooled cloud by dropping dry ice into it. He discerned correctly that dry ice, at 78.5 degrees below zero C., causes water droplets to freeze spontaneously. Within months Bernard Vonnegut, then also with General Electric, conceived the use of silver iodide as an ice-nucleating, or seeding, agent. These two discoveries provided the impetus for rain-making experiments that have been conducted in many parts of the world. After a dozen years the success of these experiments is still debated. Although it has been convincingly demonstrated that the behavior of individual clouds may be modified by seeding them from aircraft, the outcome of operations aimed at producing economically significant increases in rain over large areas is much less conclusive. Evidence is accumulating, however, that modest increases of 10 to 15 per cent may be produced in favorable circumstances.

In our laboratory we have been studying the precise conditions under which supercooled water freezes and how the freezing point may be influenced by nucleating agents of various sorts. We have found that the freezing point of water varies over a range of more than 40 degrees C., depending upon the volume of the sample, the rate of cooling and the presence of impurities that may function as nucleating agents. We have frozen many thousands of water droplets, varying in diameter from one centimeter down to a thousandth of a centimeter and all containing small foreign nuclei. The water droplets are held between layers of two liquids that are practically immiscible with water and with each other. The system is cooled at a constant rate in a refrigerator, and we record the temperature at which each drop freezes. We have found that there is a linear relationship between the freezing temperature and the logarithm of the drop diameter. Thus if one-centimeter drops of a certain sample of water freeze at 18 degrees below zero C., one-millimeter drops will freeze at 24 degrees below zero, and one-tenth-millimeter drops at 31 degrees below zero [see *illustration on page 8*]. This relationship characterizes the nucleation of water droplets by foreign particles and indicates that a decrease in temperature makes a logarithmically increasing number of atmospheric particles capable of acting as nuclei.

We have recently been successful in purifying water to such an extent that we can produce large numbers of drops entirely free of foreign particles. We accomplish this by repeatedly filtering and distilling water in a closed apparatus from which atmospheric air is rigidly excluded. One-millimeter droplets of such very pure water may be supercooled to 33 degrees below zero C. before freezing, and droplets one-thousandth of a millimeter in diameter may be cooled to 41 degrees below zero. These droplets of pure water freeze spontaneously.

Presumably small groups of water molecules, undergoing random fluctuations in position and velocity, become locked by chance into an icelike arrangement and thereby serve as nuclei to initiate the freezing process. One can calculate the rate at which such aggregates form and hence the probability that a drop of a given size will freeze. The lower curve in the illustration on page 8 indicates the computed temperatures at which droplets of various sizes should freeze within one second. The curve coincides rather well with experimental observations, and is distinctly different from the freezing curve where foreign particles play a nucleating role.

Except at very low temperatures—lower than 40 degrees below zero C.— the ice-nucleus content of the air is of fundamental importance for snow formation. It is not easily measured. A favorite method requires a cloud chamber in which a sample of atmospheric air is saturated with water vapor and rapidly

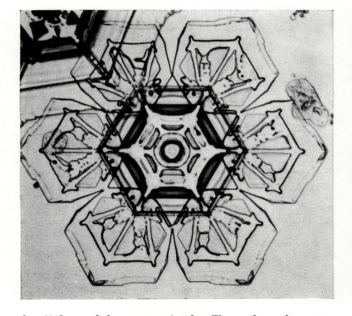

than 30 degrees below zero centigrade. (These columns happen to be formed from heavy water.) The thin hexagonal plate (*center*) is one millimeter in diameter and has petal-like extensions. Star-shaped crystal (*right*) forms the basis of the typical snowflake.

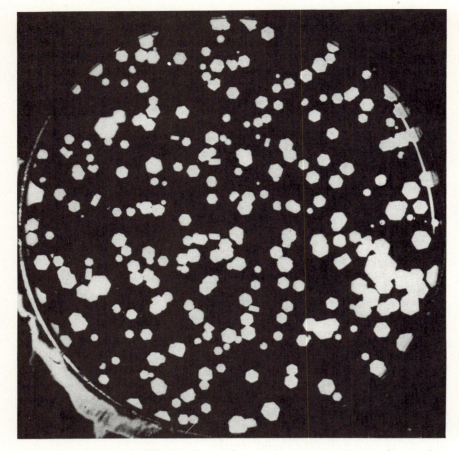

SUPERCOOLED SUGAR SOLUTION provides a way to count tiny ice crystals created in a cloud chamber. After falling on the sugar solution, crystals grow to appreciable size.

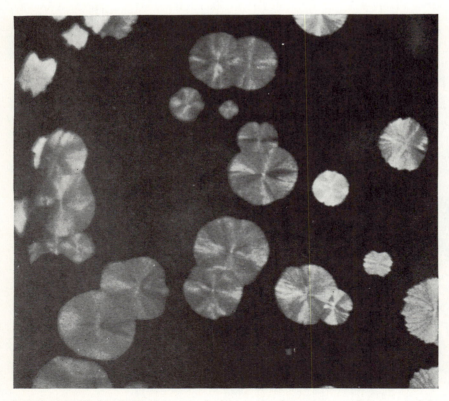

SUPERCOOLED SOAP FILM is a simple detector for determining nuclei-content of the atmosphere. Tiny ice crystals enclosing nuclei are counted after they land on film and grow.

cooled by sudden expansion. During the rapid cooling, water vapor condenses on some of the airborne particles to produce a cloud of tiny supercooled droplets. Some of these contain ice nuclei; they freeze and grow into ice crystals. The technique is then to count the number of crystals glittering in an illuminated volume of the cloud, successive measurements being made at lower and lower temperatures achieved by larger and larger expansions. Because it is not easy to discern small numbers of crystals swirling about in a thick fog, direct visual counts are not very accurate. This led my former colleague Keith Bigg to devise an ingenious technique in which the ice crystals fall into a tray of sugar solution placed at the bottom of the cloud chamber. The water in the solution supercools, and when the tiny ice crystals fall into it, they quickly grow to visible size and may be easily counted [see illustration at left].

Measurements made from aircraft over both land and sea show that the ice-nucleus population of the atmosphere varies considerably from day to day and from place to place. On some occasions it appears to fall below the minimum value required for the efficient release of precipitation from clouds. This is the justification for rain-making experiments.

The nature and origin of the nuclei necessary to initiate the formation of ice crystals are subjects of considerable interest and controversy. While I believe that they originate mainly from the earth's surface as dust particles carried aloft by the wind, E. G. Bowen of Australia's Commonwealth Scientific and Industrial Research Organization has suggested that the debris of meteorites may be an important source. He has made analyses of world rainfall patterns which seem to show some correlation with the annual meteor showers. In an attempt to test these rival hypotheses John Maybank and I have recently examined, in the laboratory, the ice-nucleating ability of various types of soil particles and mineral dust and also of meteoritic dust.

Of the 30 terrestrial dusts we have tested, 16 (mainly silicate minerals of the clay and mica variety) produced ice crystals in supercooled clouds at temperatures between 10 and 15 degrees below zero C. [see illustration on pages 12 and 13]. These substances are all minor constituents of the earth's crust. It is significant that common materials such as sea sand were not active. (Since the quartz of ordinary sand has a hexagonal

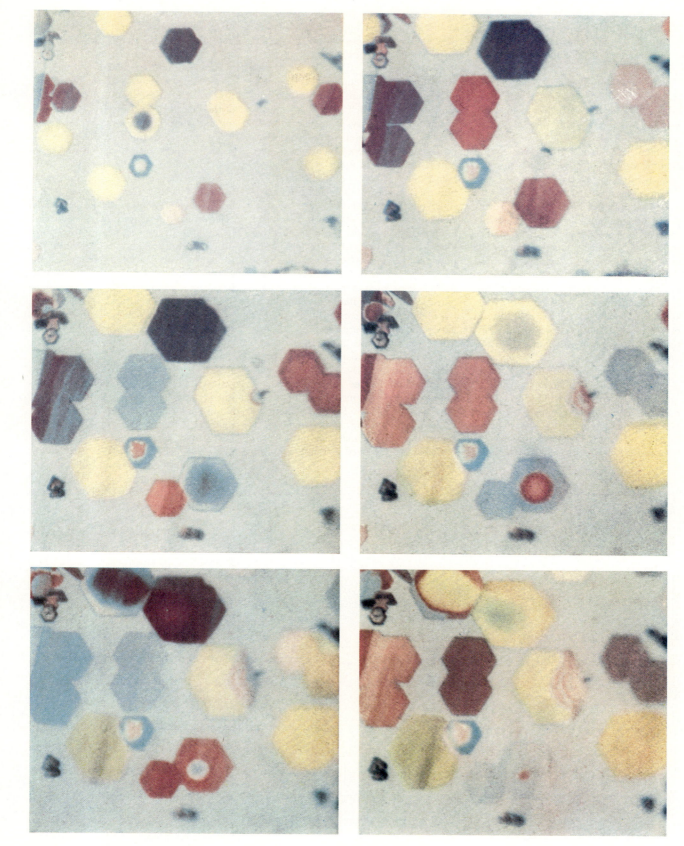

GROWTH OF SNOW CRYSTALS is revealed in this series of photomicrographs made with reflected light. The hexagonal ice crystals, growing on a single crystal of natural cupric sulfide, appear to change color (due to canceling of certain wavelengths by interference) as they become thicker. Time interval between first picture (*top left*) and second (*top right*) was 45 seconds. The rest of the series followed at 15-second intervals. Ice crystals tend to grow in diameter until they meet another crystal, then they thicken. Crystals that are of differing thickness when separate tend to acquire the same thickness after coming in contact.

crystal-structure resembling that of ice, Findeisen had thought that quartz might be an effective nucleating agent. But a superficial resemblance in structure is not enough.)

The most abundant of the active substances we have tested is kaolinite, which initiates ice formation at nine degrees below zero C. This mineral is common enough to provide an important source of ice nuclei, but not so common that the atmosphere always contains high concentrations of its particles.

These particular experiments were greatly facilitated by the use of a simple, convenient and readily renewable nucleus detector. It consists of a very stable soap film (obtained from a half-and-half mixture of water and a liquid detergent) stretched across a metal ring. When tiny ice crystals, enclosing submicroscopic nuclei, land on the supercooled soap film, they grow rapidly into crystals large enough to be easily detected and counted [see bottom illustration on page 6].

In the course of our nucleation experiments we made a surprising discovery: Ten of the terrestrial dusts were found to become more effective ice nuclei if they had previously been involved in ice-crystal formation. In other words, they could be preactivated, or "trained."

Thus when ice crystals grown on kaolinite nuclei, which are initially active at nine degrees below zero, are evaporated in a dry atmosphere, they leave behind nuclei which are thereafter effective at temperatures as much as five degrees higher. Particles of montmorillonite (another important constituent of some clays), which initially become active nuclei only at temperatures some 25 degrees below zero, may be preactivated to work at 10 degrees below zero. It seems that, although the bulk of ice surrounding the nucleus is removed during the drying process, small germs of ice, retained in pores and crevices, survive and serve as effective nuclei when the particle is again exposed to a supercooled cloud. We now have an interesting possibility. Some soil particles, such as those of montmorillonite, which are initially rather poor as ice nuclei may be carried aloft to form ice crystals at the very low temperatures associated with the high cirrus clouds. Later, if the crystals should evaporate without reaching the earth, they may leave behind trained nuclei capable of nucleating lower clouds at temperatures only a few degrees below freezing. If we accept this possibility of training initially unpromising material in the upper atmosphere, we need not interpret the

fact that efficient nuclei are occasionally more abundant at higher levels as implying that they must have entered the atmosphere from outer space.

In an attempt to provide a direct test of Bowen's meteoritic-dust hypothesis, we have tested the ice-nucleating ability of the fine dust resulting from the grinding and vaporization of several different types of stony meteorite. None has proved effective at temperatures higher than 17 degrees below zero C.

The evidence therefore appears to favor the theory that atmospheric ice-nuclei are predominantly of terrestrial origin, with the clay minerals, especially kaolinite, being a major source. Additional confirmation is provided by Japanese workers who have used the electron microscope and electron-diffraction techniques to examine the nuclei at the centers of natural snow crystals. More than three quarters of the particles were identified as soil particles, with kaolinite and montmorillonite as the most likely constituents.

Since the discovery by Vonnegut that tiny particles of silver iodide, introduced as a smoke into a supercooled cloud, cause ice crystals to appear at temperatures as high as four degrees below zero C., an intensive search has been made for other substances that might be even more effective and cheaper for cloud-seeding purposes. The table on pages 12 and 13 lists those artificial nuclei that have proved active at temperatures between four and 14 degrees below zero. The temperature shown is that at which at least one particle in 10,000 will produce an ice crystal in a supercooled cloud in the laboratory. Greater numbers of effective nuclei are obtained as the temperature is lowered below the threshold value. The first seven substances in the table are active at temperatures between four and 11 degrees, where only a very small proportion of natural ice nuclei are effective; hence the seven are all potential seeding agents. But silver iodide, being more potent and more easily dispersed than its rivals, retains first place. Nuclei of ammonium fluoride, cadmium iodide and iodine, being soluble substances, would dissolve within a minute or two of entering a water cloud, but they can be made to act as ice nuclei under special laboratory conditions.

There is a tendency for the more effective nucleators to be hexagonal crystals in which the atomic arrangement is reasonably similar to that of ice, but there are exceptions. Nevertheless for all those substances that are active above

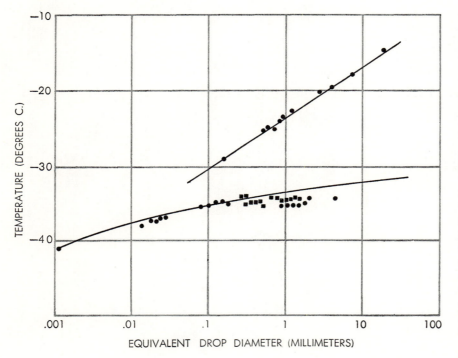

FREEZING POINT OF WATER varies with drop size. Upper curve shows freezing point of water containing impurities. Curve falls linearly with logarithmic decrease in drop size. Lower curve is theoretical freezing point for pure water and agrees well with author's values (circles) and those (squares) of Stanley Mossop of the University of Oxford.

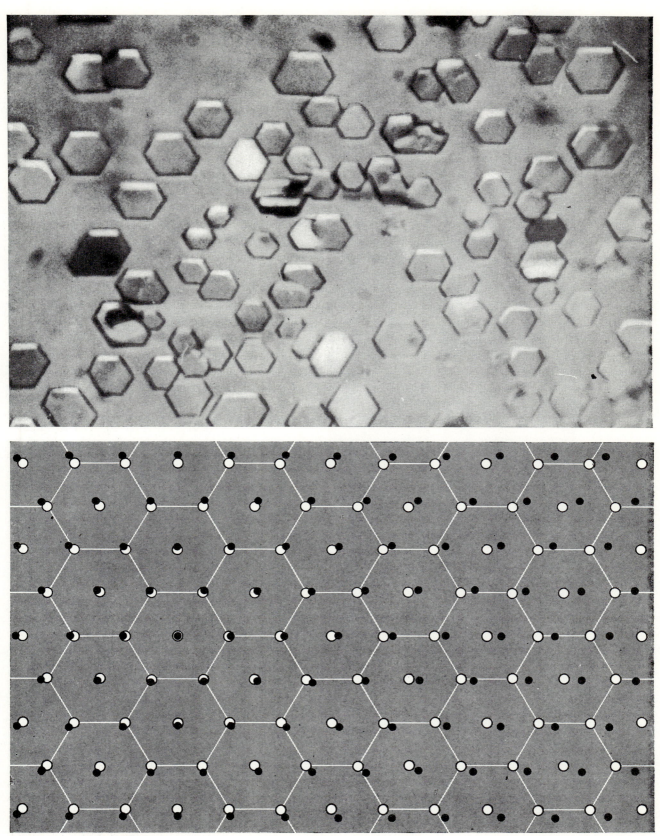

ICE GROWING ON SILVER IODIDE (*top*) shows how underlying symmetry of a large single crystal forces ice crystals to assume a parallel orientation. The relationship between the lattices of the two crystals is diagrammed below. Oxygen ions (*open circles*) define the corners of the hexagonal ice lattice; silver ions (*black dots*) lie at the corners of the silver iodide lattice. Assuming the two crystals are in perfect superposition at left of center, the match between them becomes progressively poorer in all directions, here exaggerated about threefold. Silver iodide provides best lattice fit in three dimensions of all artificial nuclei.

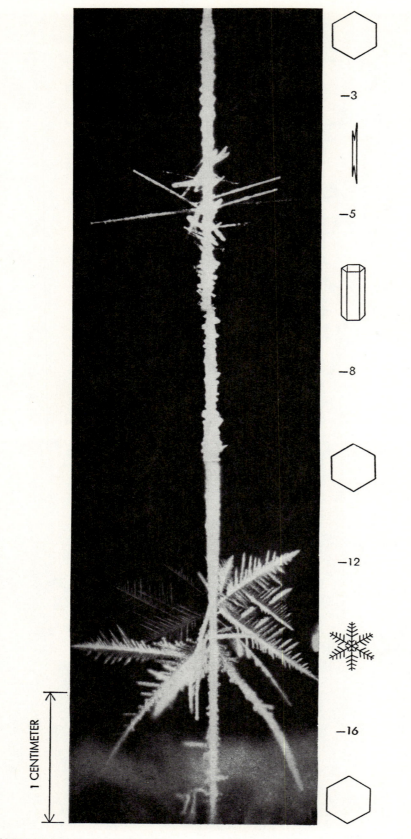

15 degrees below zero C. it is possible to find a crystal face on which the atomic spacings differ from those of ice by only a few per cent. On the other hand, we now know that the nucleating ability of a particle is not determined solely by the degree to which its atomic structure matches that of ice, and that other factors, not yet fully understood, play a role.

To investigate such factors in more detail my colleagues and I have studied, under carefully controlled conditions of temperature and humidity, the growth of ice on individual faces of single crystals of various nucleating agents. We have observed that ice crystals always assume a parallel orientation when they are grown on the hexagonal crystals of silver iodide, lead iodide, cupric sulfide, cadmium iodide and brucite, and also on crystals of calcite, mercuric iodide, iodine, vanadium pentoxide and freshly cleaved mica.

We also find that ice crystals start growing preferentially around local imperfections on the crystal surface, which show up as small dark spots in our photomicrographs. Often the ice crystals appear at the edges of steps formed during the growth or cleavage of the nucleating crystal. They also show a definite preference for the deeper steps. On lead iodide, for example, steps which are about a tenth of a micron (one tenthousandth of a millimeter) in height provide effective nucleation sites if the air is supersaturated by about 10 per cent, but much higher supersaturations, exceeding 100 per cent, are required for nucleation on the very flat areas of the host crystal.

A number of striking color effects that appear during the growth of ice crystals on a blue crystal of natural cupric sulfide (covellite) reveal much about the growth mechanism. Being only a few hundred millimicrons high, and thus comparable to the wavelength of visible light, the growing hexagonal ice plates produce interference colors that give a very accurate measure of plate thickness. The series of six photographs on page 7 are typical of many we have made. They show that in the first stages some crystals grow in diameter with no change in color, indicating no appreciable change in thickness. Evidently the water molecules arriving at the upper surface of the crystal are not captured but migrate over the surface to be built in at the crystal edges. When two crystals touch, however, they begin to thicken as colored growth-layers spread, with a speed of a few mi-

GAMUT OF ICE CRYSTAL SHAPES grows on a filament suspended in a diffusion chamber with controlled temperature gradient. Crystals take characteristic forms at various temperatures as indicated along the right edge of the photograph. Reading from the top, the symbols represent: thin hexagonal plates, needles (which are defective prisms), hollow prismatic columns, hexagonal plates, branched star-shaped crystals (or dendrites), and hexagonal plates. At temperatures lower than 25 degrees below zero C., prisms appear again.

crons per second, from the point of contact across the crystal surface. Crystals that are of differing thickness when separate tend to acquire the same thickness after coming into contact. We now know that perfect crystals grow only very slowly under normal supersaturations. It seems that to allow a crystal to grow at an observable rate, imperfections must be set up, as when a crystal accidentally hits a step or another crystal. Thus crystals sitting astride a step are often of different thickness on either side, as indicated by their two-toned appearance.

Although snow crystals occur in almost infinite variety, they can all be classified into three basic forms: the hexagonal prismatic column, the thin hexagonal plate and the branching star-shaped form, sometimes called dendritic [see illustrations on pages 4 and 5]. Until recently experts disagreed on the reasons for the differences in form. Some argued that the form depended less on temperature of formation than on the degree of supersaturation of the surrounding air.

It is almost impossible to assign reasons for the different crystal forms by collecting snowflakes on the ground. Accordingly much effort has gone into collecting ice crystals from aircraft and noting the temperature of the surrounding air. Such studies have revealed that at a given temperature a particular crystal type tends to be dominant. The high cirrus clouds, which are usually colder than 30 degrees below zero, consist of prismatic columns, typically half a millimeter long and containing pronounced funnel-shaped cavities at each end. The medium-altitude clouds, whose temperatures range between 15 and 30 degrees below zero, contain both plates and prisms. The greatest variety of crystals are found in the lower portions of supercooled clouds, where temperatures run from about five degrees below zero up to zero. Here occur hexagonal plates, short prisms, long thin needles and, most striking of all, beautiful, intricate stars up to several millimeters in diameter. Snowflakes, which are composed of from two to several hundred individual crystals, form at temperatures only a few degrees below zero.

The observation of clouds thus suggests that the shape of a snow crystal is largely controlled by the temperature of the air in which it grows. This has been confirmed in a striking manner by growing crystals under carefully controlled conditions in the laboratory. We grow crystals on a thin fiber running vertically through the center of a diffusion cloud chamber in which the vertical gradients of temperature and vapor density can be accurately controlled and measured. The results of many experiments, covering a temperature range from zero down to 50 degrees below zero and vapor supersaturations varying from a few per cent to 300 per cent, consistently show that the crystal shape varies with temperature along the length of the fiber. The cycle of shapes —from the hexagonal plates at zero degrees through needles at three degrees below zero—is always precisely reproducible. The boundaries between one form and another are very sharp [see illustration on opposite page]. For example, the transitions between plates and needles at three degrees below zero and those between hollow prisms and plates at eight degrees below zero occur within temperature intervals of less than one degree. Crystals grown from heavy water are almost identical with ordinary ice crystals except that the transition temperature between forms is shifted upward by about four degrees. This conforms with the melting point of heavy ice, 3.8 degrees above zero.

Our experiments have shown conclusively that it is temperature alone, and not supersaturation of the surrounding vapor, that governs the crystal form. The effect of supersaturation is simply to alter the growth rate: the greater the supersaturation, the faster the growth. The dependence of crystal growth upon temperature may be demonstrated in dramatic fashion merely by raising or lowering the fiber in the chamber. Whenever a crystal is thus transferred to a new environment, its further growth takes the form characteristic of the new temperature regime. By

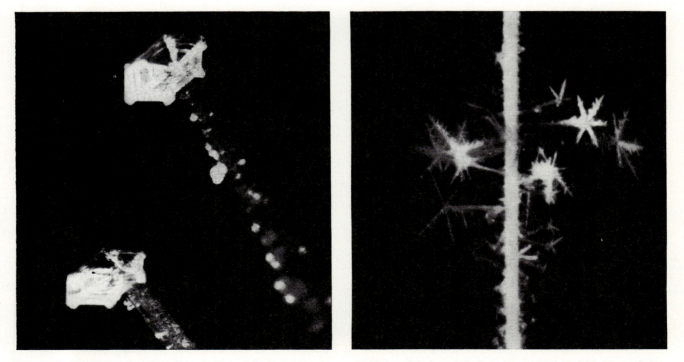

CRYSTAL HYBRIDS show how form is dictated by temperature. Needles grown at five degrees below zero C. developed plates on their ends when shifted to a temperature of 10 degrees below zero (left); stars when shifted to 14 degrees below zero (right).

thus altering the temperature we have been able to produce hybrid combinations of all the basic crystal types [see *illustrations on preceding page*].

The exact nature of the growth mechanism that can completely change the crystal shape in the space of a degree or two, and that produces five complete changes of habit in the space of only 25 degrees, is still something of a mystery. Our studies indicate, however, that the rate of migration of molecules from one crystal face to another, which appears to be very sensitive to the temperature, will be an important ingredient of the final explanation.

The work I have described does not tell us whether we can hope eventually to modify the weather, but it is aimed at establishing some of the basic physical processes that are involved in the natural formation of rain. We know that water droplets of the size found in clouds will rarely freeze spontaneously except at temperatures of 30 degrees or more below zero. We also know that there are terrestrial dusts capable of acting as nucleating agents at temperatures as high as five degrees below zero. Some

| NATURAL NUCLEI | | | | | ARTIFICIAL | |
SUBSTANCE	CRYSTAL SYMMETRY	CRYSTAL FORM	LATTICE MISFIT WITH ICE (PER CENT)	THRESHOLD TEMPERATURE (DEGREES C.)	SUBSTANCE	CRYSTAL SYMMETRY
ICE CRYSTAL	HEXAGONAL		0	0	SILVER IODIDE	HEXAGONAL
COVELLITE	HEXAGONAL		−2.8	−5	LEAD IODIDE	HEXAGONAL
BETA TRIDYMITE	HEXAGONAL		−3.5	−7	CUPRIC SULFIDE	HEXAGONAL
MAGNETITE	CUBIC		−7.1	−8	MERCURIC IODIDE	TETRAGONAL
KAOLINITE	TRICLINIC		−1.1	−9	SILVER SULFIDE	MONOCLINIC
GLACIAL DEBRIS				−10	AMMONIUM FLUORIDE	HEXAGONAL
HEMATITE	HEXAGONAL		−3.5	−10	SILVER OXIDE	CUBIC
GIBBSITE	MONOCLINIC		+12	−11	CADMIUM IODIDE	HEXAGONAL
VOLCANIC ASH				−13	VANADIUM PENTOXIDE	ORTHORHOMBIC
VERMICULITE	MONOCLINIC			−15	IODINE	ORTHORHOMBIC

NATURAL AND ARTIFICIAL ICE-NUCLEATING AGENTS tend to have hexagonal crystal habits, though there are notable exceptions. At left are nine of the 16 atmospheric dusts found to be effective nuclei at temperatures of 15 degrees below zero C. or higher. At right are 10 of the most effective artificial nuclei. The dimensional agreement between the crystal lattices of effective nuclei and the lattice of ice is usually a good one, but it does not necessarily correlate with threshold temperatures. The best fit may

of the dusts can even be trained to be more effective than they are normally. But every stable, supercooled cloud is visible evidence that effective nucleating agents are frequently lacking. Whether providing them artificially, under favorable conditions, will significantly increase snowfall or rainfall is a matter still to be resolved to the general satisfaction of meteorologists.

NUCLEI

CRYSTAL FORM	LATTICE MISFIT WITH ICE (PER CENT)	THRESHOLD TEMPERATURE (DEGREES C.)
	+1.3	−4
	+0.4	−6
	−2.8	−6
	−3.5	−8
	−0.3	−8
	−2.9	−9
	−3.8	−11
	−6.2	−12
	−2.1	−14
	−1.5	−14

occur between uneven multiples of the lattice constants of the crystal and of ice. The best fit for any one dimension of the lattices is shown in the fourth and ninth columns.

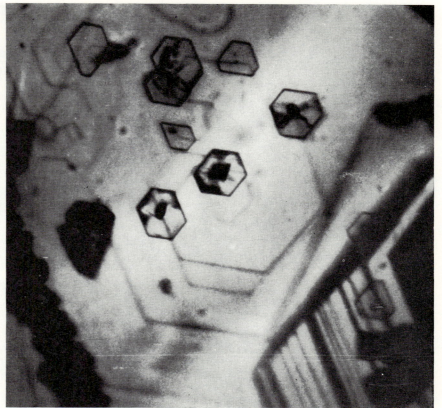

ON CADMIUM IODIDE CRYSTAL with spiral growth steps, ice crystals form preferentially at the edges of the steps. Hexagonal sides of ice crystals maintain parallelism.

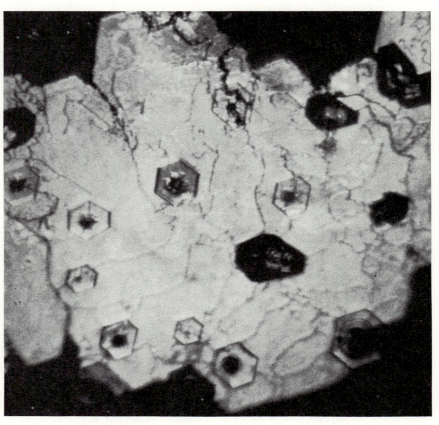

ON LEAD IODIDE CRYSTAL with a variety of growth steps resembling a contour map, ice crystals nucleate at edges of the higher steps. The most effective are about .1 micron high.

BRILLIANT COLORS displayed by several different types of aurora are captured in these photographs made by the author in the neighborhood of the University of Alaska at Fairbanks. The photograph at top left is of a quiet and diffuse auroral arc; at top right is a similar arc with a strong enhancement of the pink light emitted by excited nitrogen molecules in the earth's upper atmosphere. The remaining four photographs are of a ribbon-like auroral type known as an active rayed band. The photographs were all made with a 35-millimeter camera, using a high-speed color film and an $f/1.2$ lens. The exposure time was between one and five seconds.

The Aurora

by Syun-Ichi Akasofu
December 1965

*The information gathered by rockets and artificial
satellites has contributed to a new physical description
of the aurora in which the earth's magnetosphere acts
like a gigantic cathode-ray tube.*

It is almost impossible to capture in photographs or describe in words the unearthly beauty of the aurora as it shimmers and flames in the polar night sky. Familiar to almost everyone from pictures and descriptions but only occasionally visible where most people live, the phenomenon has long lacked a satisfactory explanation. Now, within the past few years, ground-based observations have been combined with information acquired by rockets and artificial satellites to produce a physical description of the aurora that relates it to the large-scale interaction of, on the one hand, the magnetic fields that surround the earth in space and, on the other, the high-velocity "wind" of electrically charged particles streaming from the sun. According to this view the magnetosphere of the earth acts like a gigantic cathode-ray tube that marshals charged particles into beams and focuses them on the earth's polar regions. The aurora is a shifting pattern of images displayed on the fluorescent screen provided by the atmosphere.

In more technical terms the aurora is a fluorescent luminosity produced by the interaction of atoms or molecules in the upper atmosphere and energetic charged particles entering the atmosphere from space. The incoming particles, guided by the lines of force in the earth's magnetic field, are electrons and protons. The atoms and molecules are chiefly those of oxygen and nitrogen. When they are struck by incoming particles, they are stripped of one or more electrons (ionized) or raised to a higher energy state (excited); when they return to their original condition, by acquiring electrons or by losing energy, they emit radiation of a characteristic wavelength. Thus the spectrum of the aurora can provide detailed information

about the atoms and molecules present in the upper atmosphere.

To the eye most auroras are green or blue-green, with occasional patches and fringes of pink and red. Excited oxygen atoms account for both green and red light, at the respective wavelengths of 5,577 angstrom units and 6,300 angstrom units. Ionized nitrogen molecules emit intense light, particularly violet and blue light in a group of spectral bands between 3,914 and 4,700 angstroms. Excited nitrogen molecules account for a series of emission bands that are particularly intense in the deep red part of the spectrum between 6,500 and 6,800 angstroms [*see top illustration on page 17*]. The oxygen radiation at 5,577 angstroms and the nitrogen radiation at about 3,900 angstroms originate predominantly at an altitude of about 110 kilometers (70 miles). The 6,300-angstrom radiation of oxygen originates chiefly between 200 and 400 kilometers.

At an altitude of 100 kilometers there is often no dearth of oxygen atoms excited to the level at which they could emit red light at 6,300 angstroms, but such spontaneous emission does not occur until about 200 seconds after excitation has taken place. In this period the probability is large that an excited oxygen atom will lose part of its energy in a collision with another atom or molecule. On the other hand, spontaneous emission at 5,577 angstroms (green light) takes place in about .7 second; hence green radiation predominates over red at low altitudes. Higher up in the atmosphere collisions are infrequent enough so that the 6,300-angstrom emission of excited oxygen atoms has time to take place. At such altitudes, however, the density of oxygen is so low that the red radiation is faint unless the flux of incoming particles is

high enough to excite a large fraction of all the oxygen atoms present.

Another weak source of red light is the emission of radiation by excited hydrogen atoms, which enter the atmosphere originally as protons (hydrogen nuclei). Along the way the protons pick up electrons to form hydrogen atoms. When these atoms are first created, they are in an excited state and identify themselves as they decay to lower levels by emitting the familiar Balmer series of spectral lines.

The excited and ionized states that supply most of the visible light of the aurora are produced by beams of incoming electrons that have energies of less than 10,000 volts, or less than half the energy of the electrons in the beam of a television picture tube. The energies of auroral electron beams have been measured by precisely coordinating ground-based measurements, which record luminosity profiles, and rocket or satellite measurements, which supply information on the interactions taking place in the upper atmosphere. Among those who have made important contributions to such studies are C. E. McIlwain of the University of California at San Diego, B. J. O'Brien of Rice University and a cooperating group made up of investigators at the Lockheed Aircraft Company and our laboratory at the University of Alaska.

Auroral luminosity takes two basic forms: ribbons and cloudlike patches. A vigorous auroral display generally evolves from the former to the latter, but many auroras disappear without ever breaking up into patches. A ribbon display has a vertical dimension of a few hundred kilometers and an east-west dimension of at least a few thousand kilometers. The ribbon itself is

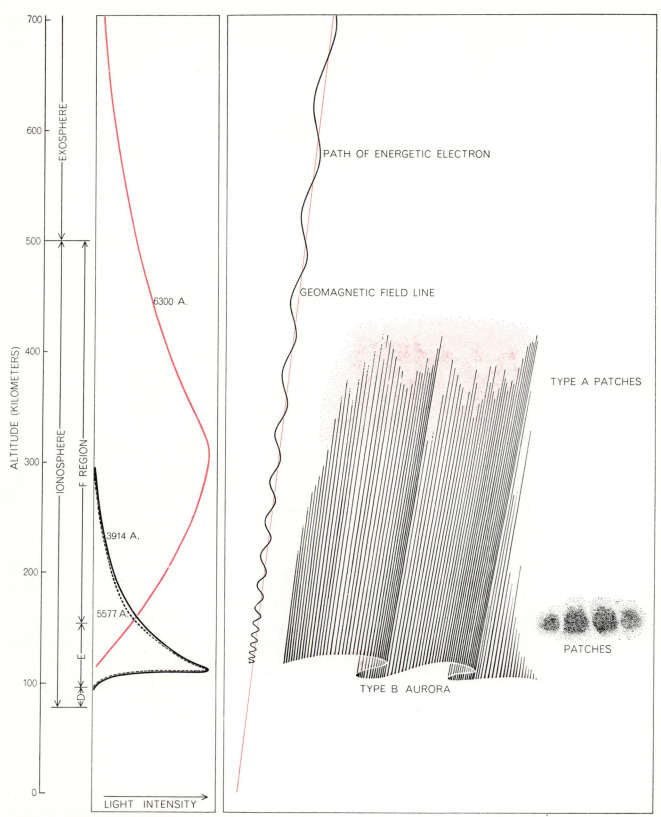

COMMON TYPES OF AURORA are shown with their average heights and characteristic radiations. The normal ribbon-like type, represented here by a "rayed band," has a vertical dimension of a few hundred kilometers and an east-west dimension of at least a few thousand kilometers. The ribbon itself is only a few hundred meters thick. An intensely active rayed band often develops a pink glow at the bottom; this is described as the Type B aurora. During the most intense activation the ribbon-like form collapses and is succeeded by cloudlike patches; these usually occur after midnight. When oxygen atoms become sufficiently excited to emit light at 6,300 angstrom units, rosy patches of Type A aurora appear at an altitude of 300 to 400 kilometers. A single energetic electron is shown gyrating down a geomagnetic-field line. Light is emitted as a result of its collisions with resident particles in upper atmosphere.

only a few hundred meters thick, which suggests that it is produced by a sheet-like electron beam that comes from the magnetosphere into the polar upper atmosphere. The ribbon form usually appears in multiple tiers—like stage curtains hanging one behind the other—stretching across the entire sky.

Students of the aurora have developed a series of descriptive terms to identify various subcategories of the ribbon form [*see bottom illustration at right*]. When the ribbon is in its simplest and quietest configuration, it is known as a homogeneous arc; at such times it has a fairly smooth luminosity, brightest at the bottom and fading into the night sky at the top. As the ribbon becomes slightly more active it develops fine folds a few kilometers in width, with the result that the aurora seems to be composed of aligned columns, or rays, of light; this is called a rayed arc. With more intense activation the folds spread to a few tens of kilometers in width. When the larger folds are superposed on the more delicate ones, the ribbon is called a rayed band. If the activation continues to increase, the rayed band develops a beautiful pink glow at the bottom of the folded ribbon; this is often described as a Type *B* aurora. Finally, if the activation rises still further, the folds or loops grow to a truly grand scale, with widths of a few hundred kilometers. As soon as the activation ceases, however, the folds tend to disappear and the ribbon resumes its homogeneous form. This suggests that the homogeneous form represents the fundamental structure of the aurora and that the folds and convolutions are indeed evidence of increased activation.

During the most intense activation the ribbon form collapses and is succeeded by the cloudlike patches; these appear most commonly after midnight. A comparison of reports from observers widely separated around the Arctic Circle leaves no doubt that the various active forms at different places and different local times are closely related to one another. For example, when one observer sees unactivated homogeneous arcs in the evening sky, other observers will report that auroras have been fairly quiet all around the Pole. On the other hand, when the quiet arcs become activated during the evening to form rayed arcs and rayed bands, observers watching the morning sky elsewhere will see previously quiet arcs break up into cloudlike patches.

I have spoken so far only of the most common types of aurora, but several

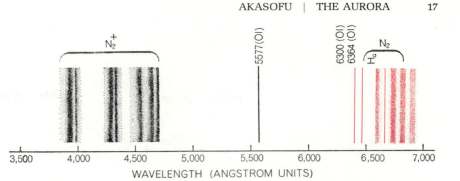

SPECTRUM OF THE AURORA can provide detailed information about the kind of particles present in the upper atmosphere and their normal energy states. Excited oxygen atoms account for both green and red light at the specific wavelengths of 5,577, 6,300 and 6,364 angstroms. Ionized nitrogen molecules supply violet and blue light in a group of spectral bands between 3,914 and 4,700 angstroms. Excited nitrogen molecules supply a series of emission bands in the deep red part of the spectrum between 6,500 and 6,800 angstroms.

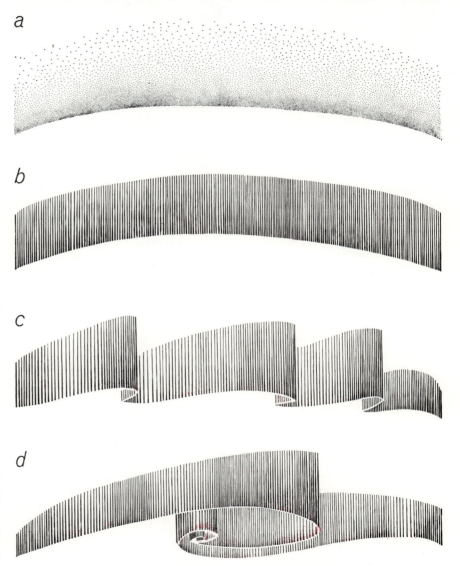

RIBBON-LIKE FORM of the aurora has various subcategories, depending on the intensity of the activation. When the ribbon is in its simplest and quietest configuration, it is known as a homogeneous arc (*a*). As the ribbon becomes slightly more active it develops fine folds and is called a rayed arc (*b*). With more intense activation larger folds are superposed on the more delicate ones and the ribbon comes to be known as a rayed band (*c*). An extremely active Type *B* aurora develops a pink glow at the bottom of the folded ribbon (*d*).

AURORAL ZONE, which is defined as a narrow band centered on the line of maximum average annual frequency of auroral visibility (*broken white line*), has only a statistical significance. At any given hour or minute slightly activated auroras tend to appear, on the average, along an oval zone that coincides with the nearly circular auroral zone only at the observer's local midnight. (In this view of the world it is approximately midnight at the Geophysical Institute of the University of Alaska at Fairbanks.) When the sun is very calm, the auroral oval is smaller and essentially circular, with its center at about the geomagnetic pole (*white dot*). As soon as the sun becomes a little active the oval expands to its average location and becomes eccentric with respect to the geomagnetic pole. During intense solar and geomagnetic storms the auroral oval shifts even farther south toward the Equator.

other varieties are seen, some only at rare intervals. One of these is the Type A aurora, a spectacular rosy variety that appears at an altitude of 300 to 400 kilometers when oxygen atoms become sufficiently excited to emit light at 6,300 angstroms. Type A auroras occur only a few times in a dozen years. Incoming protons frequently give rise to an extensive but faint band of whitish-green luminosity. During intense magnetic storms a "midlatitude red arc" appears at latitudes considerably south of New York City; it consists of a wide but not very bright band of 6,300-angstrom radiation and thus is not easily seen with the unaided eye. A few hours after an intense solar flare the entire polar region is bombarded by energetic protons expelled from the sun, producing the whitish-green "polar-cap glow." In addition to all these natural varieties of

aurora, high-altitude nuclear-bomb tests have produced brilliant crimson auroras. None of these special types of aurora need concern us further.

What are the actual mechanisms involved in the production of the aurora? Instruments carried by satellites have shown that the earth and its magnetic field are confined in a huge cavity carved in the solar wind. The cavity is the magnetosphere; it contains all the belts of particles trapped in the earth's magnetic field [see "The Solar Wind," by E. N. Parker, SCIENTIFIC AMERICAN, April, 1964, and "The Magnetosphere" by Laurence J. Cahill, Jr., SCIENTIFIC AMERICAN, March, 1965].

Furthermore, the magnetosphere has a long cylindrical tail that is carried downstream by the solar wind; this was first suggested by J. H. Piddington of the

Commonwealth Scientific and Industrial Research Organization in Australia. Norman Ness of the Goddard Space Flight Center has recently confirmed the existence of this comet-like tail with satellite measurements. Within the cylindrical tail the lines of force in the earth's magnetic field are bunched together like a bundle of spears. The lines of force above the plane of the magnetosphere's equator are directed toward the sun; those below the plane are directed away from the sun. The equatorial plane therefore constitutes a neutral sheet. The secret of the aurora seems to be hidden in the tail of the magnetosphere.

The role of the magnetosphere's tail can be visualized by comparing the magnetosphere to a cathode-ray tube; this analogy was suggested to me by C. T. Elvey, formerly director of the Geophysical Institute of the University of Alaska. In a cathode-ray tube electrons are emitted by a heated filament and accelerated toward an anode that is perforated with a small hole. Some of the electrons pass through the hole and form a pencil-like beam that is deflected, on its passage to the face of the tube, by electric fields between two pairs of plates or, in some tubes, by magnetic fields set up by coils. The electron beam strikes a fluorescent material on the tube face, producing a luminous image. The luminous display on the screen thus supplies evidence of changes in both electric and magnetic fields along the path of the electron beam.

In much the same way the shifting patterns of the aurora over the entire polar night sky supply evidence of changes in the magnetic and electric fields along the path of electrons streaming toward the earth. In some obscure fashion the tail of the magnetosphere accelerates and collimates electrons in the magnetosphere into ribbon-like beams that impinge sharply on the upper atmosphere. The task ahead is to identify the precise mechanisms that play the role of the electron gun, the anode, the electric plates and the magnetic coils.

Clues to these mechanisms can be found in changes in the magnetic field on the earth's surface and in space, as well as in the auroral display itself. Let us, therefore, "watch" the auroral display as it appears on the "screen" of the entire polar night sky. If one travels northward from the border between the U.S. and Canada, one will see the aurora with increasing frequency. The increase does not continue all the way to the North Pole; the frequency reaches a maximum over the southern part of Hudson Bay. Excellent maps have been prepared that show auroral "isochasms": the lines of equal average annual frequency of visible auroras. The auroral zone is defined as a narrow band centered on the line of the maximum isochasm. There is, of course, an auroral zone in the Southern Hemisphere as well as in the Northern. The center of the northern auroral zone is not, as one might think, the magnetic dip pole near Resolute Bay in Canada (73.5 degrees north latitude and 100 degrees west longitude), but what is known as the dipole, or geomagnetic pole, at the northwestern tip of Greenland (78.5 degrees north and 69 degrees west).

The auroral zone, however, has only

a statistical significance. At any given hour or minute auroras tend to appear, on the average, in an oval zone that coincides with the nearly circular auroral zone only at the observer's local midnight [see illustration on opposite page]. Elsewhere the oval zone falls inside the auroral zone. The oval zone, if it were viewed from a point above the geomagnetic pole, would appear as an oval glow roughly fixed in space above the geomagnetic pole. The earth turns below the oval pattern once a day, and the locus of the midnight portion of the oval traces out a circle that coincides with the auroral zone. The auroral oval is a new concept that has evolved gradually as the result of cooperation among workers in Australia, Canada, Denmark, Finland, Norway, Sweden, the U.S.S.R., the United Kingdom and the U.S.

From investigations of the outer belt of particles trapped in the earth's magnetic field (particularly studies by James A. Van Allen and L. Frank of the State University of Iowa), it seems likely that the oval belt of the aurora lies immediately poleward of the curve of intersection between the ionosphere (which begins about 80 kilometers above the earth's surface) and the shell of trapped particles that forms the outer boundary of the inner magnetosphere [see illustration below]. This region is populated with electrons whose energies range upward from about 40,000 volts. Such electrons are produced in the tail of the magnetosphere and flow along the boundary of the inner magnetosphere, creating a more or less steady glow where they intersect with the upper atmosphere. The oval has the shape it

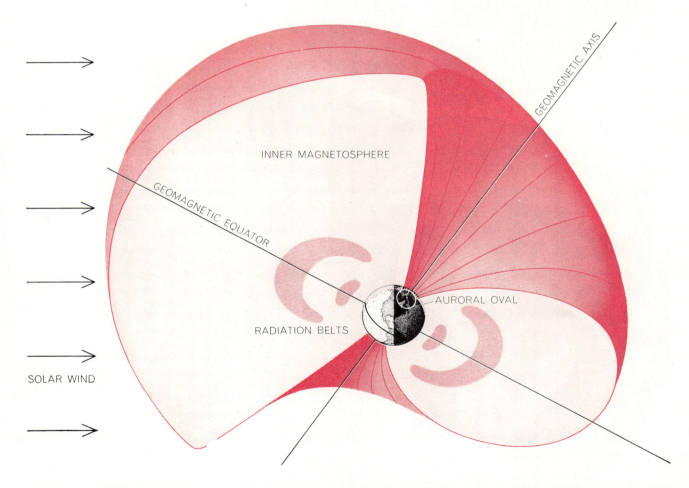

THE MAGNETOSPHERE is a vast cavity in the solar wind that contains the earth's magnetic field; its role in the production of the aurora is to act like a gigantic cathode-ray tube that focuses electron beams, as well as proton beams, toward the earth in the vicinity of the magnetic poles. The oval belt of the aurora coincides roughly with the intersection curve between the ionosphere, which begins about 80 kilometers above the earth's surface, and the outer boundary of the inner magnetosphere (color). This region of the magnetosphere is populated with electrons whose energy ranges from about 40,000 volts upward; the extended tail of the magnetosphere, whose magnetic-field lines terminate inside the auroral ovals, is not shown here. The auroral oval has the shape it does because the inner magnetosphere is highly asymmetric with respect to the solar wind, and hence to the earth's noon-midnight axis.

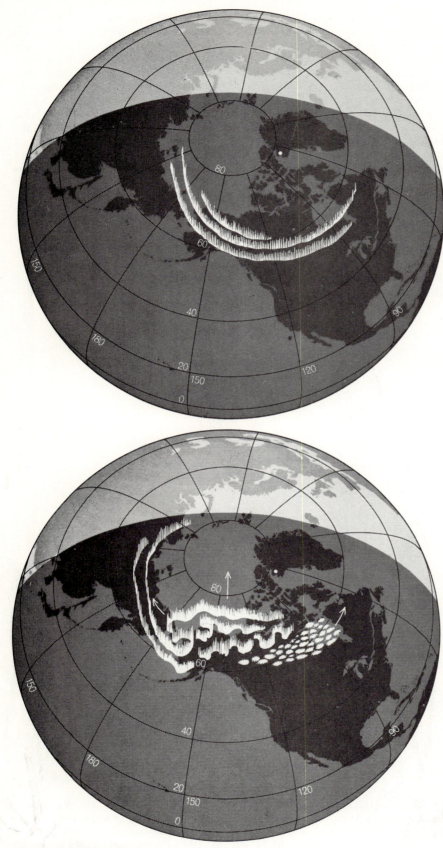

AURORAL SUBSTORM, the most spectacular auroral display, occurs during major magnetic storms and coincides with a rapid buildup of circumpolar ring currents. Such substorms originate in the midnight sector of the auroral oval and usually last for two or three hours. These two views show the quiet homogeneous arcs of the aurora before the substorm (*top*) and the complex display of many auroral types at the substorm's height (*bottom*).

does because the magnetosphere is highly asymmetric with respect to the solar wind and hence to the noon-midnight axis on the earth's surface.

The size of the oval changes greatly with solar activity. When the sun is extremely quiet, the midnight portion of the oval recedes toward the geomagnetic pole and the oval becomes almost circular and faint. The oval occupies its typical position when the sun is in a state of average activity. After a major solar storm the oval expands rapidly toward the Equator, the amount of expansion being roughly proportional to the volume of electric current flowing in the tail of the magnetosphere and also to the intensity of a gigantic ring current that grows within the magnetosphere and flows westward around the earth. The expansion of the oval toward the Equator indicates that the inner magnetosphere is shrinking. Thus by watching changes in the oval one can visualize large-scale changes in the internal structure of the magnetosphere.

The most dynamic auroral displays occur during major magnetic storms, which are evoked in turn by intense solar activity. The most violent displays, known as auroral substorms, coincide with a rapid buildup of the circumpolar ring currents in the ionosphere. Such substorms originate in the midnight sector of the auroral oval and usually last for two or three hours [*see illustration at left*]. The first indication of the substorm is a sudden increase in the brightness of one of the quiet arcs. Soon the brightened auroras, having assumed the character of rayed arcs and bands, begin to spread explosively toward the poles at speeds of about five kilometers per second or even higher. When the poleward expansion is very rapid, the rayed bands begin breaking up in the midnight sector.

Meanwhile, in the evening sector of the polar region, the auroral expansion generates large-scale folds and loops that travel westward along preexisting arcs with a speed of about a kilometer per second. Such westward surges commonly sweep across the Alaskan sky and then the Siberian sky only 20 or 30 minutes after they have first appeared in the Canadian sky.

In the morning sector of the polar region the auroral arcs or bands that lie in the northern part of the oval also expand explosively, but those that lie in the southern part of the oval often disintegrate into cloudlike patches. Both the bright bands and the patches drift rapidly eastward. As the substorm subsides, the scattered auroral frag-

MOST INTENSE AURORAL SUBSTORM of recent years occurred during the great magnetic storm of February 11, 1958, which resulted from an intense solar flare on February 9. By 1020 hours (universal time) on February 11 at least three large substorms were observed. At the height of the magnetic storm (*left*) the auroral oval was driven so far south that it formed a line connecting Red- mond, Ore., Vermillion, S.D., Williams Bay, Wis., Ithaca, N.Y., and Hanover, N.H. Curiously no bright auroras could be seen north of this line. Then a fourth substorm began and within 30 minutes the active auroras had spread northward until they covered a band some 2,000 kilometers wide (*right*). An intense shower of X rays was recorded at the height of the fourth substorm.

ments converge slowly and reassemble into quiet homogeneous arcs, tracing out the auroral oval as it was before the substorm began.

Perhaps the most intense auroral substorm of recent years occurred during the great magnetic storm of February 11, 1958, which resulted from an intense solar flare on February 9. About 29 hours after the flare the magnetosphere was enclosed in a violent wind of charged particles. Within a few hours ring currents began building up within the magnetosphere and reached maximum intensity between 1000 and 1100 hours (universal time) on February 11. During this buildup period at least three large auroral substorms occurred. At the height of the magnetic storm the auroral oval was driven so far south that it formed a line connecting Redmond,

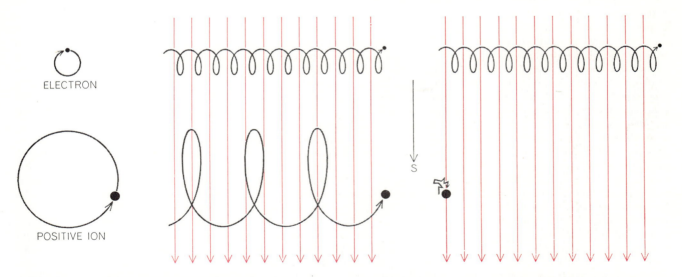

ELECTRONS AND POSITIVE IONS in the earth's ionosphere respond differently to the onset of an auroral substorm, depending on their altitude. Before such a substorm (*left*) electrons are forced to rotate clockwise, whereas positive ions are forced to rotate counterclockwise, around the geomagnetic-field lines, which in this view must be imagined as entering the page at right angles from above. The events that underlie the substorm also give rise to an electric field (*colored arrows*) that points southward, causing both electrons and positive ions to be deflected eastward. At levels above the E region of the ionosphere (*middle*) the density of gas particles is so slight that electrons and positive ions have about equal mobility; even though both may travel violently eastward together there is no net displacement of electric charge and hence no flow of current. In the E region (*right*), however, the positive ions, being larger, collide more frequently with other particles than the electrons do and thus tend to be stopped, whereas the electrons continue to drift eastward. This separation of electric charges gives rise to the westward-flowing polar electrojet (*see illustration on next page*).

POLAR ELECTROJET (*black arrows*), an intense electric current that flows westward around the auroral oval, occurs only during an auroral substorm. Part of the electrojet leaks away from the auroral oval and spreads southward over the whole ionosphere (*white arrows*).

Ore., Vermillion, S.D., Williams Bay, Wis., Ithaca, N.Y., and Hanover, N.H. Curiously no bright auroras could be seen north of this line. Then the fourth substorm began, and within 30 minutes the active auroras had spread northward until they covered a band some 2,000 kilometers wide [*see top illustration on preceding page*]. In anticipation of the magnetic storm J. R. Winckler and his co-workers at the University of Minnesota sent a balloon aloft over Minneapolis with radiation-recording instruments. At the height of the fourth substorm the instruments recorded an intense shower of X rays generated when energetic auroral electrons collided with atoms and molecules in the upper atmosphere.

X-ray production is just one of several interesting phenomena associated with the auroral substorm. For example, during such a storm intense electron beams penetrate deeper into the atmosphere than usual and greatly increase the ionization of the lower parts of the ionosphere known as the E

and D regions. The E region reflects short radio waves and makes long-distance radio communication possible. When abnormal ionization occurs in the D region, communication is often disrupted. On the other hand, heavy ionization in that region tends to scatter shorter radio waves: the microwaves that ordinarily pass through the ionosphere. When such waves are scattered, television viewers sometimes receive a channel that originates in a distant city. It has also been observed that the atmosphere is heated and expands upward during the auroral substorm, causing satellites to decelerate at a slightly faster rate.

The large-scale activity of the magnetosphere during the substorm has at least one other major effect. When the magnetosphere interacts with the E region of the ionosphere, it generates the "polar electrojet," an intense electric current that flows westward around the auroral oval [*see illustration above*]. To understand the origin of this current, let us imagine that we are looking down on the polar ionosphere, where we find

that the lines of force in the earth's magnetic field are penetrating the ionosphere from above. Under the influence of this field electrons moving through the upper ionosphere are forced to travel in a clockwise direction, whereas positive ions are forced to travel in a counterclockwise one [*see bottom illustration on preceding page*].

When energetic charged particles flow out from the tail of the magnetosphere toward the earth, they develop (in the Northern Hemisphere) an electric field that points southward. Such a field, being directed at right angles to the magnetic field, causes both electrons and positive ions to be deflected eastward. At levels above the E region the density of gas particles is so low that electrons and positive ions have about equal mobility; even though both may flow strongly eastward together, there is no significant net displacement of electric charge and hence no flow of current. In the E region, however, the positive ions, being larger, collide more frequently with other particles than the electrons do and thus tend to be stopped, whereas the electrons continue to drift eastward. This separation of electric charges gives rise to the westward polar electrojet. (According to convention, current flow is opposite to electron flow.) It has been found that part of the electrojet leaks away from the auroral oval and spreads southward over the whole ionosphere.

The study of the auroral substorm owes much to photographic records made by an all-sky camera, the first models of which were designed by Carl W. Gartlein of Cornell University. This instrument consists of a camera mounted above a convex mirror that reflects the whole sky from horizon to horizon. The all-sky camera was first used extensively during the International Geophysical Year (1956–1957), when 67 nations participated in a worldwide geophysical program. Since then all-sky photographs from as many as 115 stations in the Arctic and Antarctic have been collected and studied by our laboratory at the University of Alaska.

Our task has been to analyze successive photographs of the auroral substorm taken simultaneously at various polar stations and to infer minute-to-minute changes in the magnetosphere. These analyses and inferences are then correlated whenever possible with measurements made simultaneously by instrumented satellites. Out of this coordinated study has emerged the account of the aurora I have presented in this article.

The Shape of Raindrops

by James E. McDonald
February 1954

*They are not handsomely tapered but often resemble
a small hamburger bun. This unpoetical form, frozen
by high-speed photography, is analyzed to reveal the
forces that mold it.*

One of the few predictions that a meteorologist can make with confidence of almost 100 per cent accuracy is that if you ask an illustrator or cartoonist to draw a falling raindrop, his picture will be dead wrong. Without fail he will give it the streamlined shape commonly known as the teardrop. Meteorologists have known for many years that a real raindrop bears no resemblance to this drawing-board impostor. As pictured by high-speed photography, the usual small-sized drop (less than a millimeter in diameter) is almost perfectly spherical, and a larger drop is a squat object resembling nothing so much as a hamburger bun.

While this real picture is esthetically less satisfying than the teardrop fiction, it is of considerable interest to meteorologists. Just why the large raindrops take the deformed shape that they do has been a puzzle for half a century. I became interested in the problem in a casual way which seems to illustrate how the non-systematic approach in science can sometimes bear fruit. While browsing through a work on surface physics, I came upon a general equation which described the internal pressure in any fluid object, whatever the shape of its surface curvature. A slightly different equation, applying only to a spherical shape, had previously been tried on the drop problem without complete success. This more generalized relation was immediately recognizable as the key to an understanding of the deformation of large raindrops. It took only a few minutes to formulate the general outlines of a new hypothesis for raindrop shape. But to obtain an adequate check on this new hypothesis and to expand it in certain necessary details required several months' work. The results seem to show

that one can, in fact, give a fairly good account of the way in which oversized raindrops develop their hamburger-bun shape. They also bring out that a falling raindrop is the seat of some surprisingly complex physical processes.

One might begin by asking why a drop of water should bear any resemblance to a sphere at all. The answer is that surface tension always tends to reduce the surface of a free mass of liquid to the smallest area it can achieve. The smallest possible surface area is that of a sphere, and an isolated drop of liquid not distorted by external forces is pulled by its surface tension into a spherical shape. In terms of thermodynamics, it adjusts itself to the spherical shape to minimize its surface free energy.

The question can also be considered in terms of pressure. The internal pressure just inside the convex surface of a drop is higher than the external pressure prevailing in the surrounding gas. The smaller the radius of curvature, the greater is the pressure difference between the two sides of the convex surface. (For example, a cloud droplet of one micron radius has an internal pressure of more than two atmospheres.) If our isolated drop should momentarily assume some shape other than that of a sphere, its surface would have different radii of curvature at different points, and the internal pressure just below the surface would momentarily be dissimilar at these various points. The consequent pressure gradients within the drop would tend to force liquid from the regions of sharp surface curvature to those of more gentle curvature. This is equivalent to saying that surface tension, through its control of internal pressure, reshapes the drop into a sphere when-

ever it happens to become slightly deformed. When the drop is finally brought into the shape of a perfect sphere, the uniform surface curvature makes the pressure difference uniform at all points of its surface, and the internal pressure within the drop also is uniform, provided that the external pressure field around the drop remains so.

Such a uniform pressure field actually exists in an ordinary fog or cloud, and, sure enough, photomicrographs of cloud and fog drops show that these tiny particles are indistinguishable from perfect spheres. Doubtless precise measurements would disclose that they depart slightly from perfect sphericity, because of the weak gravitational and aerodynamic forces acting on them, but a manufacturer of precision ball-bearings would be very happy to turn out bearings as close to perfection as these drops.

The disturbing effects on drop shapes that appear as one considers larger and larger drops seem to be due almost entirely to aerodynamic and gravitational forces. A. F. Spilhaus, now at the University of Minnesota, was the first to call attention to the role played by aerodynamic forces in shaping raindrops. As drops grow larger and their falling speed increases, the disturbance of the air sets up momentary non-uniformities in the air pressure around them. The big drops that pelt down in a summer thunderstorm plummet through the air so fast that they continuously create about them their own shape-deforming pressure fields as they fall.

Now it is well known that when a body falls the air pressure just under the body becomes higher than average and the pressure around its sides lower than average. This means, of course, that the internal pressures within a large falling

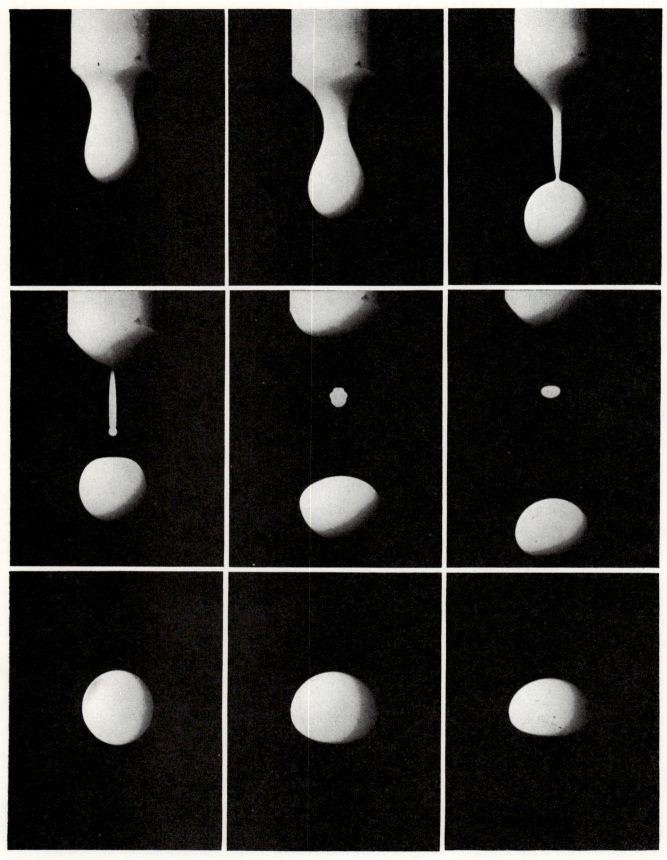

FALLING DROPS OF MILK, used instead of water because of their high visibility, were photographed with high-speed flash by Harold E. Edgerton of the Massachusetts Institute of Technology. In the third, fourth and fifth pictures the shape of the drop briefly oscillates. In the sixth it is spherical. In the last picture the drop has attained an almost stable configuration after a fall of 14 feet.

raindrop must change accordingly; the drop develops an excess of pressure near its bottom and a deficiency of pressure all around its waist. And if we momentarily assume that the air flow is that of a perfect fluid, there will be an excess of internal pressure near the top of the drop as well as near the bottom.

At this point we obtain a vivid notion of how a drop takes care of its own shape. The gradients of internal pressure drive water from near the base and top out into the regions around the waist, thereby tending to flatten the drop and increase its horizontal diameter. Even more intriguing is the fact that the resultant modification of the drop's surface curvature is of just the kind required to help the drop restore a uniform internal pressure and achieve an equilibrium. The sharpening of the curvature around the waist adjusts the surface tension effects to make up for the deficiency of external pressure there, while the flattening of curvature near the base and top tends to cancel the effects of the excessive external pressures in those regions. Together the joint action of surface tension and aerodynamic forces deforms the drop continuously until it reaches a stable internal pressure distribution.

But clever as a raindrop may now seem in managing its affairs, one must ask whether a large drop has truly brought itself into complete mechanical equilibrium when its internal pressure is uniform. The answer is that it has not, for it must still meet an important demand of the laws of gravity: that is, it must develop a vertical pressure gradient just sufficient to permit the lower strata of the drop to hold up the upper ones in the gravitational field. Briefly, a liquid drop falling at terminal velocity can be in full mechanical equilibrium only when its internal pressure, instead of being uniform throughout, varies vertically in such a way as to satisfy the familiar hydrostatic equation relating liquid density, liquid depth and the acceleration of gravity. If the drop were accelerating freely in the earth's gravitational field, this hydrostatic requirement would not appear. But raindrops reach a terminal uniform velocity after only a few yards of fall; hence small but important hydrostatic pressure gradients must exist within them.

In the tiny droplets of clouds the hydrostatic gradients are not important, because the difference in internal pressure from top to bottom of these microscopic globules is negligibly small compared to their internal pressure increase

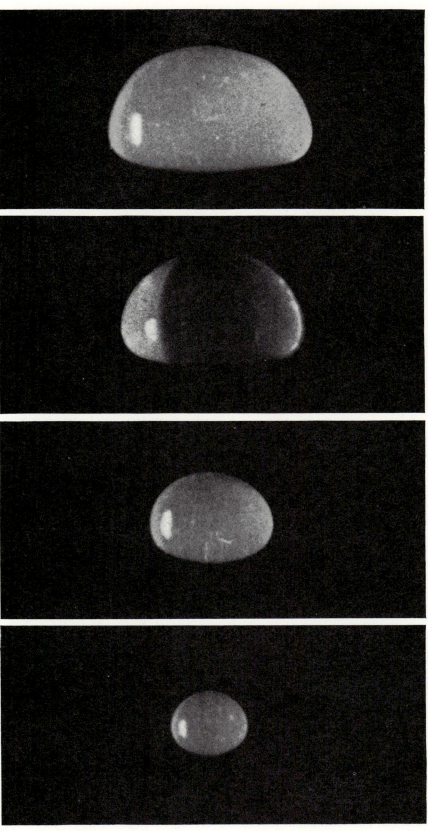

FALLING DROPS OF WATER were photographed by Choji Magono of Yokohama National University. The volume of the drop at the top is equivalent to that of a sphere 6.5 millimeters in diameter; its velocity is 8.9 meters per second. The corresponding numbers for the other drops are as follows. Second drop: 6 mm. and 8.8 meters per second. Third: 4.8 mm. and 8.3 meters per second. Fourth: 2.8 mm. and 6.8 meters per second.

due to surface tension. The top-to-bottom hydrostatic pressure differential varies directly with drop size, while the surface pressure increment varies inversely with drop size. Consequently, by the time a cloud droplet has grown into a large raindrop hydrostatic effects have become about equal in importance to surface-tension effects—a point which appears to have been overlooked by those who have examined the raindrop-shape problem in the past.

Combining the hydrostatic principles with the aerodynamic principles, one next obtains a curious result. If we demand that drops deform in such a manner as to yield an internal pressure field satisfying the hydrostatic equation, we find to our embarrassment that the drop must be flatter on its upper side than on its lower—which is just the reverse of the shape that falling raindrops actually take.

An aerodynamicist would spot the difficulty very quickly. We have assumed so far that the drops are falling through a perfect, non-viscous fluid. Actually air has some viscosity, enough to have a significant effect on an object of raindrop size (one to five millimeters in diameter) and falling speed (five to eight meters per second). Around a large raindrop the air must behave essentially as it does over the wing of an airplane in a stall: the boundary layer of air just next to the object (raindrop or aircraft wing) separates from the object and leaves a turbulent wake. In such a wake region the air pressure is always lower than it would be for perfect fluid flow in which the streamlines neatly close in behind the object. Thus the lower air pressure in the wake of a falling drop forces a greater curvature of its upper surface than of its underside.

When the drop has accomplished this adjustment, it is at last in full equilibrium with all of the important forces that play upon it as it cleaves through the air. It seems almost unfair that the fate of so cleverly equilibrated a little system may be no more glorious than to splatter down on some dusty road at the beginning of an August thundershower.

The role of boundary-layer separation came to light only after I had computed pressure profiles from measurements made on an actual photograph of water drops provided by Choji Magono, now of Yokohama National University. The method used to deduce the aerodynamic pressure distribution over the surface of a drop hinged upon the use of certain relationships concerning the

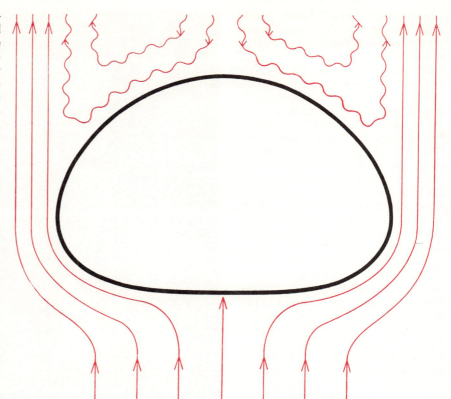

FLOW OF AIR around a large falling raindrop is indicated by the red lines in this diagram. The boundary layer streamlines follow the curve of the drop until they reach the "separation point." Above the drop and enclosed by the separating boundary layer is a turbulent region. The low pressure of this region is responsible for the shape of the drop.

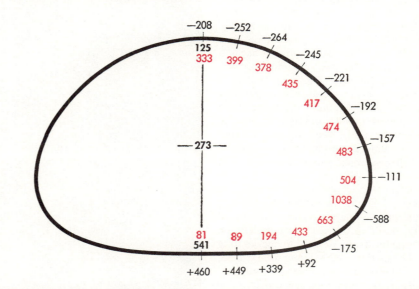

DISTRIBUTION OF PRESSURE is given for a 6-mm. drop falling 8 meters per second. The numbers outside the drop give the difference between the pressure around the drop and that of the atmosphere. The red numbers inside give the difference in surface pressure; the black numbers, in internal pressure. The units are dynes per square centimeter.

differential geometry of surfaces of revolution obligingly worked out by my colleague J. M. Keller at Iowa State College. The falling speed and size of the photographed drop were known, and its curvature at every point could be accu-

rately measured on the picture. The first step in the analysis was to determine the pressure at a single point. Fortunately the pressure at the center of the underside of a falling body can be calculated even if nothing is known about the rest of the airflow pattern. This, added to the pressure due to surface tension which was computed from the curvature at the bottom, gave the total pressure just inside the lowest point of the drop. Since the drop was in hydrostatic equilibrium, the internal pressures at all the other levels could be calculated. The surface tension pressures at other points were calculated from curvature measurements and subtracted from the internal pressure. This gave the outside air pressure all around the drop. The pressure pattern thus deduced turned out to be much like the patterns observed in wind-tunnel work on separating boundary layers.

The idea that there was a separation of boundary layers around raindrops agreed with observations. Ross Gunn of the U. S. Weather Bureau had reported a curious sideslipping of falling drops of about one millimeter diameter, and he had shown rather convincingly that this odd behavior must involve eddies. Ed-

dies can be shed only from a wake region enclosed by a separating boundary layer. Later a more quantitative type of evidence for the shape hypothesis was obtained by means of a calculation of the total pressure drag acting on a falling drop. In the case of a drop falling with uniform velocity, this pressure drag plus the drag of friction (which is very small) must be equal to the weight of the drop. It was found that the drag was in fact equal to the weight to within the limits of precision of the methods employed in the pressure calculation. It now appears reasonably safe to conclude that the queer shape of a large raindrop results in an understandable way from a conspiracy among the forces we have here examined.

Of what use is the result? First of all, it is always pleasant to acquire some understanding of even a minor peculiarity of nature. Then also, the study yield-

ed information which may be useful for solving the vexed problem of why and how raindrops break into fragments in the turbulent regions of clouds. Finally, the clear recognition of the role of boundary-layer separation in the aerodynamics of raindrops will almost certainly help to clarify the nature of heat and vapor transport at the surface of falling raindrops.

But, to end on a note of dark pessimism, it seems quite improbable that nny amount of progress in exploring the drop-shape problem will persuade cartoonists and commercial artists to alter the shape of their peculiar brand of raindrops.

4

River Meanders

by Luna B. Leopold and W. B. Langbein
June 1966

The striking geometric regularity of a winding river is no accident. Meanders appear to be the form in which a river does the least work in turning; hence they are the most probable form a river can take.

Is there such a thing as a straight river? Almost anyone can think of a river that is more or less straight for a certain distance, but it is unlikely that the straight portion is either very straight or very long. In fact, it is almost certain that the distance any river is straight does not exceed 10 times its width at that point.

The sinuosity of river channels is clearly apparent in maps and aerial photographs, where the successive curves of a river often appear to have a certain regularity. In many instances the repeating pattern of curves is so pronounced that it is the most distinctive characteristic of the river. Such curves are called meanders, after a winding stream in Turkey known in ancient Greek times as the Maiandros and today as the Menderes. The nearly geometric regularity of river meanders has attracted the interest of geologists for many years, and at the U.S. Geological Survey we have devoted considerable study to the problem of understanding the general mechanism that underlies the phenomenon. In brief, we have found that meanders are not mere accidents of nature but the form in which a river does the least work in turning, and hence are the most probable form a river can take.

Regular Forms from Random Processes

Nature of course provides many opportunities for a river to change direction. Local irregularities in the bounding medium as well as the chance emplacement of boulders, fallen trees, blocks of sod, plugs of clay and other obstacles can and do divert many rivers from a straight course. Although local irregularities are a sufficient reason for a river's not being straight, however, they are not a necessary reason. For one thing, such irregularities cannot account for the rather consistent geometry of meanders. Moreover, laboratory studies indicate that streams meander even in "ideal," or highly regular, mediums [see *illustration on page 32*].

That the irregularity of the medium has little to do with the formation of meanders is further demonstrated by the fact that meandering streams have been observed in several naturally homogeneous mediums. Two examples are ocean currents (notably the Gulf Stream) and water channels on the surface of a glacier. The meanders in both cases are as regular and irregular as river meanders.

The fact that local irregularities cannot account for the existence of river meanders does not rule out other random processes as a possible explanation. Chance may be involved in subtler and more continuous ways, for example in turbulent flow, in the manner in which the riverbed and banks are formed, or in the interaction of the flow and the bed. As it turns out, chance operating at this level can explain the formation of regular meanders. It is a paradox of nature that such random processes can produce regular forms, and that regular processes often produce random forms.

Meanders commonly form in alluvium (water-deposited material, usually unconsolidated), but even when they occur in other mediums they are invariably formed by a continuous process of erosion, transportation and deposition of the material that composes the medium. In every case material is eroded from the concave portion of a meander, transported downstream and deposited on the convex portion, or bar, of a meander. The material is often deposited on the same side of the stream from which it was eroded. The conditions in which meanders will be formed in rivers can be stated rather simply, albeit only in a general way: Meanders will usually appear wherever the river traverses a gentle slope in a medium consisting of fine-grained material that is easily eroded and transported but has sufficient cohesiveness to provide firm banks.

A given series of meanders tends to have a constant ratio between the wavelength of the curve and the radius of curvature. The appearance of regularity depends in part on how constant this ratio is. In the two drawings on page 30 the value of this ratio for the meander that looks rather like a sine wave (*top*) is five for the wavelength to one for the radius; the more tightly looped meander (*bottom*) has a corresponding value of three to one. A sample of 50 typical meanders on many different rivers and streams has yielded an average value for this ratio of about 4.7 to one. Another property that is used to describe meanders is sinuosity, or tightness of bend, which is expressed as the ratio of the length of the channel in a given curve to the wavelength of the curve. For the large majority of meandering rivers the value of this ratio ranges between 1.3 to one and four to one.

Close inspection of the photographs

ENTRENCHED MEANDERS of the Colorado River in southern Utah were photographed from a height of about 3,000 feet. The meanders were probably formed on the surface of a gently sloping floodplain at about the time the entire Colorado Plateau began to rise at least a million years ago. The meanders later became more developed as river cut deep into layers of sediment. Mean downstream direction is toward right.

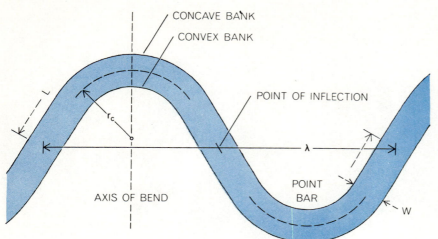

CONCAVE BANK
CONVEX BANK
POINT OF INFLECTION
AXIS OF BEND
POINT BAR
λ
r_c
L
W

WIDTH OF CHANNEL (W) = 1
WAVELENGTH (λ) = 11.5
LENGTH OF CHANNEL (L) = 16.5
RADIUS OF CURVATURE (r_c) = 2.3

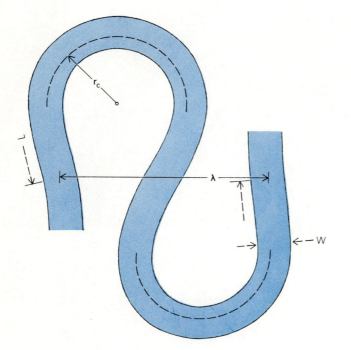

r_c
L
λ
W

WIDTH OF CHANNEL (W) = 1
WAVELENGTH (λ) = 6.9
LENGTH OF CHANNEL (L) = 24.8
RADIUS OF CURVATURE (r_c) = 2.3

PROPERTIES used to describe river meanders are indicated for two typical meander curves. A series of meanders has a regular appearance on a map whenever there tends to be a constant ratio between the wavelength (λ) of the curve and its radius of curvature (r_c). The value of this ratio for the meander that looks rather like a sine wave (top) is five to one; the more tightly looped meander (bottom) has a corresponding value of three to one. An average value for this ratio is about 4.7 to one. Sinuosity, or tightness of bend, is expressed as the ratio of the length of the channel (L) in a given curve to the wavelength of curve. The value of this ratio for the top curve is 1.4 to one and for the bottom curve 3.6 to one. On the average the value of this ratio ranges between 1.3 to one and four to one.

and maps that accompany this article will show that typical river meanders do not exactly follow any of the familiar curves of elementary geometry. The portion of the meander near the axis of bend (the center of the curve) does resemble the arc of a circle, but only approximately. Neither is the curve of a meander quite a sine wave. Generally the circular segment in the bend is too long to be well described by a sine wave. The straight segment at the point of inflection—the point where the curvature of the channel changes direction—prevents a meander from being simply a series of connected semicircles.

Sine-generated Curves

We first recognized the principal characteristics of the actual curve traced out by a typical river meander in the course of a mathematical analysis aimed at generating meander-like curves by means of "random walk" techniques. A random walk is a path described by successive moves on a surface (for example a sheet of graph paper); each move is generally a fixed unit of distance, but the direction of any move is determined by some random process (for example the turn of a card, the throw of a die or the sequence of a table of random numbers). Depending on the purpose of the experiment, there is usually at least one constraint placed on the direction of the move. In our random-walk study one of the constraints we adopted was that the path was to begin at some point A and end at some other point B in a given number of steps. In other words, the end points and the length of the path were fixed but the path itself was "free."

The mathematics involved in finding the average, or most probable, path taken by a random walk of fixed length had been worked out in 1951 by Hermann von Schelling of the General Electric Company. The exact solution is expressed by an elliptic integral, but in our case a sufficiently accurate approximation states that the most probable geometry for a river is one in which the angular direction of the channel at any point with respect to the mean down-valley direction is a sine function of the distance measured along the channel [see illustration on opposite page].

The curve that is traced out by this most probable random walk between two points in a river valley we named a "sine-generated" curve. As it happens, this curve closely approximates the shape of real river meanders [*see illustration on next page*]. At the axis of bend the channel is directed in the mean down-valley direction and the angle of deflection is zero, whereas at the point of inflection the angle of deflection reaches a maximum value.

A sine-generated curve differs from a sine curve, from a series of connected semicircles or from any other familiar geometric curve in that it has the smallest variation of the changes of direction. This means that when the changes in direction are tabulated for a given distance along several hypothetical meanders, the sums of the squares of these changes will be less for a sine-generated curve than for any other regular curve of the same length. This operation was performed for four different curves of the same length, wavelength and sinuosity—a parabolic curve, a sine curve, a circular curve and a sine-generated curve—in the illustration on page 33. When the squares of the changes in direction were measured in degrees over 10 equally spaced intervals for each curve, the resulting values were: parabolic curve, 5,210; sine curve, 5,200; circular curve, 4,840; sine-generated curve, 3,940.

Curve of Minimum Total Work

Another property closely associated with the fact that a sine-generated curve minimizes the sum of the squares of the changes in direction is that it is also the curve of minimum total work in bending. This property can be demonstrated by bending a thin strip of spring steel into various configurations by holding the strip firmly at two points and allowing the length between the fixed points to assume an unconstrained shape [*see top illustration on pages 34 and 35*]. The strip will naturally avoid any concentration of bending and will assume a shape in which the bend is as uniform as possible. In effect the strip will assume a shape that minimizes total work, since the work done in each element of length is proportional to the square of its angular deflection. The shapes assumed by the strip are sine-generated curves and indeed are good models of river meanders.

A catastrophic example of a sine-generated curve on a much larger scale was provided by the wreck of a Southern Railway freight train near Greenville, S.C., on May 31, 1965 [*see bottom illustration on page 35*]. Thirty adjacent flatcars carried as their load 700-foot sections of track rail chained in

a bundle to the car beds. The train, pulled by five locomotives, collided with a bulldozer and was derailed. The violent compressive strain folded the trainload of rails into a drastically foreshortened snakelike configuration. The elastic properties of the steel rails tended to minimize total bending exactly as in the case of the spring-steel strip, and as a result the wrecked train assumed the shape of a sine-generated curve that distributed the bending as uniformly as

possible. This example is particularly appropriate to our discussion of river meanders because, like river meanders, the bent rails deviate in a random way from the perfect symmetry of a sine-generated curve while preserving its essential form.

The Shaping Mechanism

The mechanism for changing the course of a river channel is contained

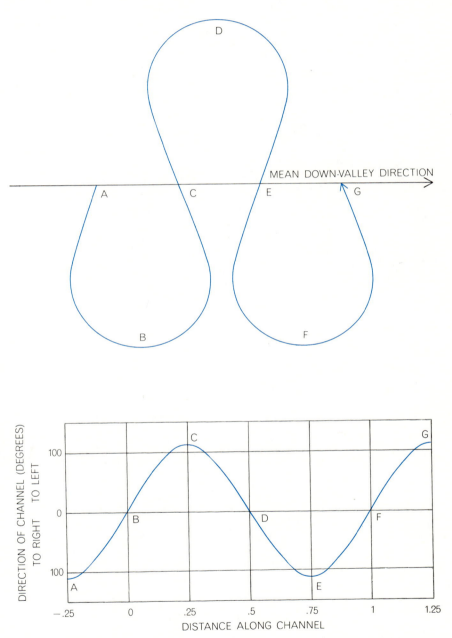

SINE-GENERATED CURVE (*top*) closely approximates the shape of real river meanders. This means that the angular direction of the channel at any point with respect to the mean down-valley direction (*toward the right*) is a sine function of the distance measured along the channel (*graph at bottom*). At the axis of each bend (**B, D and F**) the channel is directed in the mean down-valley direction and the angle of deflection is zero, whereas at each point of inflection (**A, C, E and G**) the angle of deflection reaches a maximum value.

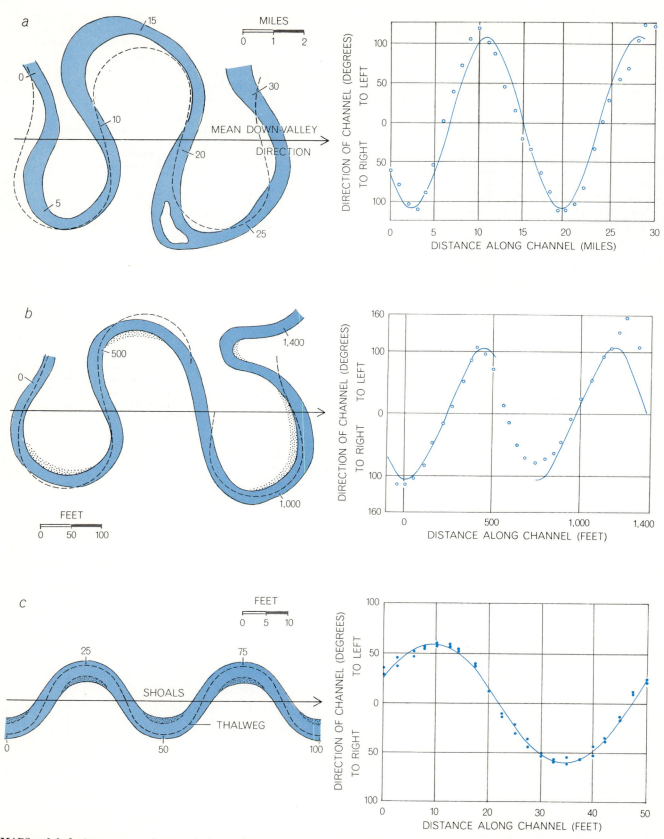

MAPS at left depict segments of two typical meandering streams, the Mississippi River near Greenville, Miss. (*a*), and Blackrock Creek in Wyoming (*b*), as well as a segment of an experimental meander formed in a homogeneous medium in the laboratory (*c*). Measurements of the angular direction of the channels with respect to the mean downstream direction were made at regular inter- vals along the center lines of the two natural meanders and along the thalweg, or deepest part of the channel, of the experimental meander. When these measurements were plotted against the dis- tance of each channel, the resulting curves closely approximated sine waves (*right*). The corresponding sine-generated curves are superposed on their respective channel maps (*broken black curves*).

in the ability of water to erode, transport and deposit the material of the river's medium. Especially on a curve, the velocity gradient against the channel bank sets up local eddies that concentrate the expenditure of energy and localize erosion. An idealized flow pattern in a typical meander is shown in the top illustration on page 36. The left side of the illustration indicates the velocity vectors at various points for five cross sections along the curve. As the cross sections indicate, the depth of the channel changes systematically along the curve, the shallowest section being at the point of inflection and the deepest section at the axis of bend. At the same time the cross-sectional shape itself changes; it is symmetrical across the channel just downstream from the point of inflection and most asymmetrical at the axis of bend, the deeper section being always nearer the concave bank. The velocity vectors show a normal decrease in velocity with depth except at the axis of bend and near the concave bank, where the highest velocity at any point in the meander occurs somewhat below the surface of the water.

The right side of the same illustration shows the streamlines of flow at the surface of the meander. The maximum-velocity streamline is in the middle of the channel just downstream from the point of inflection; it crosses toward the concave bank at the axis of bend and continues to hug the concave bank past the next point of inflection. Riverboatmen navigating upstream on a large river face the problem that the deepest water, which they usually prefer, tends to coincide with the streamline of highest velocity. Their solution is to follow the thalweg (the deepest part of the river, from the German for "valley way") where it crosses over the center line of the channel as the channel changes its direction of curvature but to cut as close to the convex bank as possible in order to avoid the highest velocity near the concave bank. This practice led to the use of the term "crossover" as a synonym for the point of inflection.

The lack of identity between the maximum-velocity streamline and the center line of the channel arises from the centrifugal force exerted on the water as it flows around the curve. The centrifugal force is larger on the faster-moving water near the surface than on the slower-moving water near the bed. Thus in a meander the surface water is deflected toward the concave bank, requiring the bed water to move toward the convex bank. A circulatory system

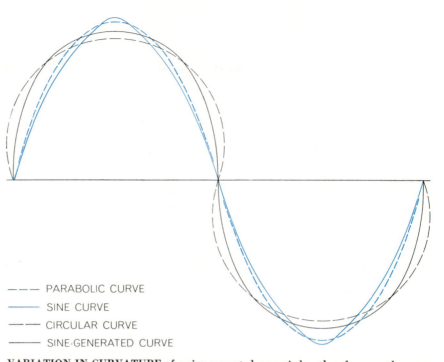

--- PARABOLIC CURVE

—— SINE CURVE

—- — CIRCULAR CURVE

——— SINE-GENERATED CURVE

VARIATION IN CURVATURE of a sine-generated curve is less than for any other regular geometric curve. This means that when the changes in direction are tabulated for small distances along several hypothetical meanders, the sums of the squares of the changes in direction will be less for a sine-generated curve than for any other curve. The changes in direction were measured in degrees over 10 equally spaced intervals for each of the four curves depicted here. When the squares of these changes were summed, the following values were obtained: parabolic curve, 5,210; sine curve, 5,200; circular curve, 4,840; sine-generated curve, 3,940. The four curves are equal in length, wavelength and sinuosity.

is set up in the cross-sectional plane, with surface water plunging toward the bed near the concave bank and bed water rising toward the surface near the convex bank. This circulation, together with the general downstream motion, gives each discrete element of water a roughly helical path that reverses its direction of rotation with each successive meander. As a result of this helical motion of water, material eroded from the concave bank tends to be swept toward the convex bank, where it is deposited, forming what is called a point bar.

Erosion of the concave banks and deposition on the convex banks tends to make meander curves move laterally across the river valley. Because of the randomness of the entire process, the channel as a whole does not move steadily in any one direction, but the combined lateral migration of the meanders over a period of many years results in the river channel's occupying every possible position between the valley walls. The deposition on the point bars, combined with the successive occupation by the river of all possible positions, results in the formation of the familiar broad, flat floor of river valleys—

the "floodplain" of the river. The construction of a floodplain by the lateral movement of a single meander can be observed even in the course of a few years; this is demonstrated in the bottom illustration on page 36, which is made up of four successive cross sections surveyed between 1953 and 1964 on Watts Branch, a small tributary of the Potomac River near Washington.

The overall geometry of a meandering river is an important factor in determining the rate at which its banks will be eroded. In general the banks are eroded at a rate that is proportional to the degree with which the river channel is bent. Any curve other than a sine-generated curve would tend to concentrate bank erosion locally or, by increasing the total angular bending, would add to the total erosion. Thus the sine-generated curve assumed by most meandering rivers tends to minimize total erosion.

Riffles and Pools

In the light of the preceding discussion it is possible to examine some of the hydraulic properties of meanders in greater detail. If a river channel is re-

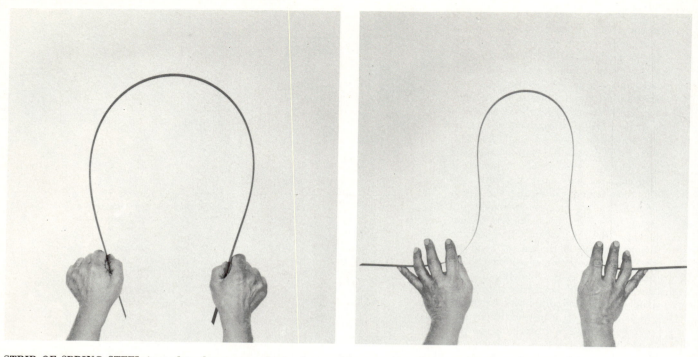

STRIP OF SPRING STEEL is used to demonstrate that a sine-generated curve is the curve of minimum total work. The strip is bent into various configurations by holding it firmly at two points and allowing the length between the fixed points to assume an un-

garded as being in a steady state, the form it assumes should be such as to avoid concentrating variations in *any* property at the expense of another property.

For example, variations in depth and velocity are inherent in all river channels, whether they are straight or curved. Even a reach, or length of channel, that is quite straight has a more or less uneven bed that consists of alternating deeps and shallows. Although this is not so obvious in a period of high flow, it becomes quite apparent at low flow, when the shallow sections tend to ripple in the sunlight as water backs up behind each hump in the bed before pouring over its downstream slope. To a trout fisherman this fast reach is known as a riffle. Alternating with the riffles are deeps, which the fisherman would call pools, through which the water flows slower and more smoothly.

The alternation of riffles and pools in a trout stream at low flow is noteworthy for another reason. The humps in the stream bed that give rise to the riffles tend to be located alternately on each side of the stream [*see top illustration on page 37*]. As a consequence the stream at low flow seems to follow a course that wanders successively from one side of the channel to the other, in a manner having an obvious similarity to meandering.

The analogy between this temporary sinuosity and full-scale meandering is strengthened by the fact that the riffles occur at roughly equal intervals along the channel. Moreover, the spacing of the riffles is correlated with the width of the channel. Successive riffles are located at intervals equal to about five to seven times the local channel width, or roughly twice the wavelength of a typical meander. This surprisingly consistent ratio seems even more remarkable when one realizes that each meander contains two riffles, one at each point of inflection. This observation led us to hypothesize that the same mechanism that causes meanders must also be at work in straight channels, and that a detailed study of the form and the hydraulic properties of two segments of channel that differ only in their degree of curvature might shed some light on the formation of meanders.

Obtaining Meander Profiles

In order to test this hypothesis it was necessary to obtain accurate data for all the pertinent hydraulic factors: depth, velocity, water-surface profile and bed profile. For several years we had attempted to measure such factors in small rivers near Washington just after every heavy rainstorm, when there was a rapid increase in streamflow. The water level changed so quickly in such storms, however, that there was never enough time to measure all the hydrau-

lic factors in detail through a succession of two riffles and an intervening pool. Then in 1959 we tried another strategy: we decided to measure a small stream in Wyoming, named Baldwin Creek, in early June, a period of maximum runoff from melting snow. Measurements were made in two places, a meandering reach and a straight reach, that were comparable in all outward aspects except sinuosity. The stream was about 20 feet wide and was nearly overflowing its banks, so that we could just barely walk in it wearing chest-high rubber waders.

Robert M. Myrick, an engineer with the Geological Survey, and one of us (Leopold) began a series of measurements in the midafternoon of June 19, surveying water-surface and bed profiles with a level and a rod, and making velocity and depth measurements with a current meter and a rod. When darkness came, we lighted lanterns and continued our measurements. At about daybreak we slept for a few hours and then resumed the survey, grateful that the melting snow had kept the stream at a steady high flow for such a long time.

Several days later we were able to sit down under a tree and plot the profiles, velocities and depths on graph paper. What emerged was a quite unexpected contrast between meandering reach and straight reach [*see bottom illustration on page 37*]. The slope of the water surface in the meandering reach

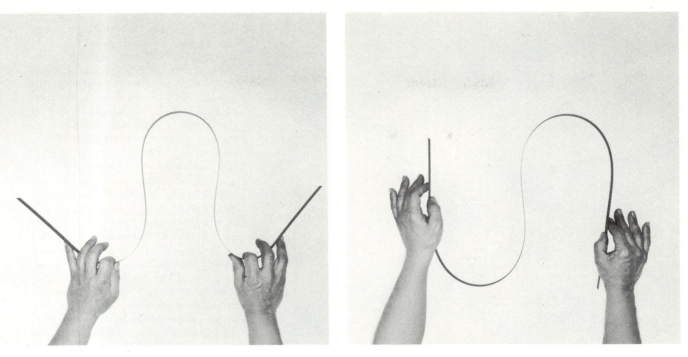

constrained shape. The strip will naturally avoid any concentration of bending and will assume a shape in which the bend is as uniform as possible. In each of the four cases shown here this shape is a sine-generated curve and indeed a good model of a river meander.

was clearly steeper than that in the straight reach; moreover, the water-surface profile of the meandering reach was nearly a straight sloping line, whereas the straight reach had a stepped profile, steep over the riffle bars and comparatively flat over the intervening pools.

What did this mean? It was as if the river had, to use somewhat anthropomorphic terms, chosen to cut a meander curve in order to achieve a more uniform water-surface profile. This suggested that the river had chosen the curved path in order to achieve the objective of uniform energy loss for each unit of distance along the channel, but had paid a price in terms of the larger total energy loss inherent in a curved path.

Conclusions

These data provided the key to further research, which ultimately resulted in several conclusions. First, it appears that a meandering channel more

CATASTROPHIC EXAMPLE of a sine-generated curve on a much larger scale was provided by the wreck of a Southern Railway freight train near Greenville, S.C., on May 31, 1965. Thirty adjacent flatcars carried as their load 700-foot sections of track rails chained in a bundle to the car beds. The train, pulled by five locomotives, collided with a bulldozer and was derailed. The violent compressive strain folded the trainload of rails into the drastically foreshortened configuration shown in this aerial photograph. The elastic properties of the steel rails tended to minimize total bending exactly as in the case of the spring-steel strip shown at top of these two pages, and the wrecked train assumed the shape of a sine-generated curve that distributed the bending as uniformly as possible.

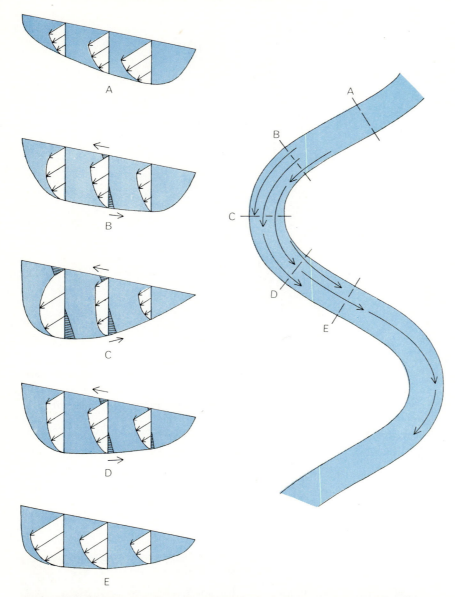

IDEALIZED FLOW PATTERN of a typical meander is shown here. The left side of the illustration indicates the velocity vectors in a downstream direction for five cross sections across the curve; the lateral component of the velocity is indicated by the triangular hatched areas. The right side of the illustration shows the streamlines at the surface of the meander.

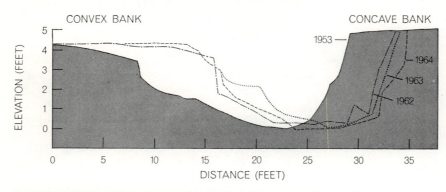

LATERAL MIGRATION of a typical meander is demonstrated in this drawing, made up of four successive cross sections surveyed between 1953 and 1964 on Watts Branch, a small tributary of the Potomac River near Washington. The lateral migration of meanders by the erosion of the concave banks and deposition on the convex banks over many years results in a river channel's occupying every possible position between the valley walls.

closely approaches uniformity in the rate of work over the various irregularities of the riverbed than a straight channel does. Of course the slope of the water surface is, with a slight correction for velocity, an accurate indicator of the rate at which energy is lost in the form of frictional heat along the length of the stream. Therefore a uniform longitudinal water-surface slope signifies a uniform expenditure of energy for each unit of distance along the channel.

A meander attains a more uniform rate of energy loss by the introduction of a form of energy loss not present in a straight reach, namely the curved path. It is evident that work is required to change the direction of a flowing liquid. Thus the slope of the water surface should increase wherever a curve is encountered by a river. In a meander it is at the deep pools, where the water-surface slope would be less steep than the average, that the introduction of a curve inserts enough energy loss to steepen the slope, thereby tending to make the slope for each unit of river length nearly the same. Accordingly the alternation of straight shallow reaches with curved deep reaches in a meander appears to be the closest possible approach to a configuration that results in uniform energy expenditure.

It is now possible to say something about the development of meandering in rivers. Although one can construct in a laboratory an initially straight channel that will in time develop a meandering pattern, a real meandering river should not be thought of as having an "origin." Instead we think of a river as having a heritage. When a continent first emerges from the ocean, small rills must form almost immediately; thereafter they change progressively in response to the interaction of uplift and other processes, including irregularities in the hardness of the rock.

Today the continuous changes that occur in rivers are primarily wrought by the erosion and deposition of sedimentary material. As we have seen, rivers tend to avoid concentrating these processes in any one place. Hence any irregularity in the slope of a river—for example a waterfall or a lake—is temporary on a geological time scale; the hydraulic forces at work in the river tend to eliminate such concentrations of change.

The formation of meander curves of a particular shape is an instance of this adjustment process. The meandering form is the most probable result of the processes that on the one hand tend to

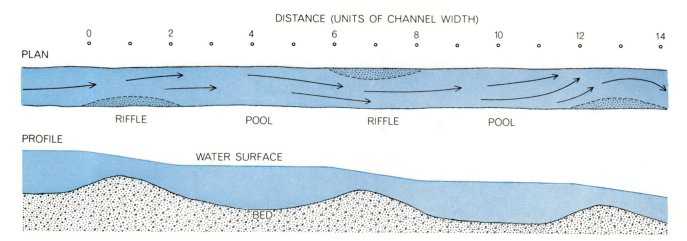

DISTANCE (UNITS OF CHANNEL WIDTH)

PLAN

RIFFLE   POOL   RIFFLE   POOL

PROFILE

WATER SURFACE

"BED"

STRAIGHT REACH of a river has a more or less uneven bed that consists of alternating deeps and shallows, known to trout fishermen as riffles and pools. The humps in the stream bed that give rise to the riffles tend to be located alternately on each side of the stream at intervals roughly equal to five to seven times the local stream width. As a consequence the stream at low flow seems to follow a course that wanders from one side of the channel to the other, in a manner having an obvious similarity to meandering.

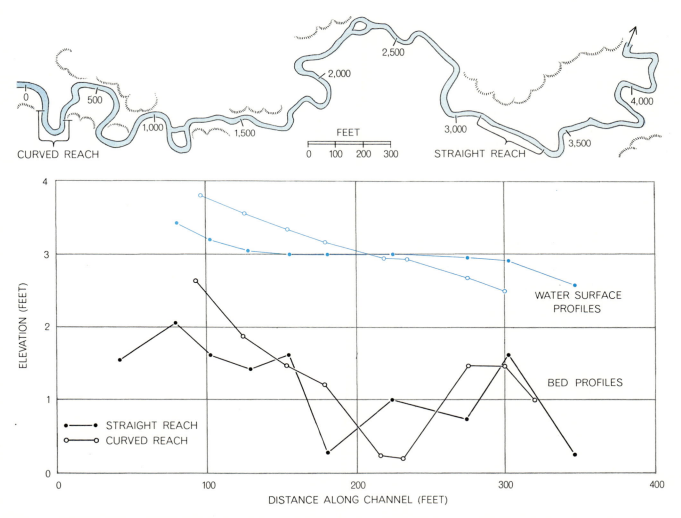

PROFILES of the water surface and bed of a small stream in Wyoming named Baldwin Creek were obtained by one of the authors (Leopold) and a colleague in 1959 during a period of maximum runoff from melting snow. Measurements were made in two places, a meandering reach and a straight reach, that were comparable in all outward aspects except sinuosity (*map at top*). What emerged was a quite unexpected contrast between the two reaches (*bottom*). The slope of the water surface in the meandering reach was clearly steeper than that in the straight reach; moreover, the water-surface profile of the meandering reach was nearly a straight sloping line, whereas the straight reach had a stepped profile, steep over the riffle bars and comparatively flat over the intervening pools.

eliminate concentrations of energy loss and on the other tend to reduce the total energy loss to a minimum rate. The sine-generated curvature assumed by meanders achieves these ends more satisfactorily than any other shape.

The same tendencies operate through the erosion-deposition mechanism both in the river system as a whole and in a given segment of the river. The tendency toward uniform power expenditure in the entire river leads toward a longitudinal profile of the river that is highly concave, inasmuch as uniformity in the rate of work per unit of length of channel would be achieved by concentrating the steepest slopes near the headwaters, where the tributaries and hence discharges are small. The longitudinal concavity of the river's profile also minimizes work in the system as a whole.

Such a longitudinal concave profile, however, would lead to considerable variation in the rate of energy expenditure over each unit area of channel bed. Uniformity in this rate would be best achieved by a longitudinal profile that was nearly straight rather than by one that was highly concave. Actual river profiles lie between these two extremes, and meanders must be considered in both contexts: first, as they occur within the river system as a whole, and second, as they occur in a given segment of channel.

In the context of the entire river system a meander will occur where the material constituting the banks is comparatively uniform. This will be more likely to take place downstream in a floodplain area than upstream in a headwater area. To the extent that the meandering pattern tends to lengthen the downstream reaches more than those upstream, it promotes concavity in the longitudinal profile of the system, thereby promoting uniformity in the rate of energy expenditure per unit of channel length.

In the local context of a given segment of channel the average slope of the channel is fixed by the relation of that segment to the whole profile. Any local change in the channel must maintain that average slope. Between any two points on a valley floor, however, a variety of paths are possible, any one of which would maintain the same slope and hence the same length. The typical meander shape is assumed because, in the absence of any other constraints, the sine-generated curve is the most probable path of a fixed length between two fixed points.

Thunder

by Arthur A. Few
July 1975

*It is the acoustic signal generated by a rapidly
expanding channel of heated air. From the information
in this signal it is possible to deduce the location,
shape and orientation of a lightning flash.*

It is obvious today that the ultimate cause of thunder is lightning; it is less obvious just how an electrical discharge in the atmosphere produces the variety of powerful noises heard in a thunderstorm. A lightning flash dissipates a prodigious amount of energy, but how is that energy (or a portion of it) transformed into sound waves? Moreover, a lightning stroke is essentially instantaneous; how does it generate the often protracted sequence of rumbles, claps, booms and other sounds heard in a peal of thunder?

Considering that thunder has been a topic of speculation since antiquity, one might expect that these questions would have been answered long ago. Actually even the most general principles of thunder production were not established until this century, and a theory of thunder that is both comprehensive and detailed has been approached only in the past 10 or 15 years. A number of important questions remain unresolved today.

With my colleagues at Rice University and in cooperation with workers elsewhere I have investigated thunder by recording its "signature": the pattern of sound waves received at a particular location from a single lightning flash. The study of such signatures has revealed much about the nature of thunder itself, and it has led to the formulation of a theory of how features of the lightning discharge are expressed in the acoustic signal. The theory is successful enough for us now to employ thunder as a tool in searching for the source of electricity in clouds.

The relation of lightning to thunder was well established by the end of the 19th century. It remained to be determined what effects of the lightning discharge are responsible for producing the sound. Four principal theories were proposed, and in the first decade of the 20th century they were vigorously debated. (In 1903 *Scientific American* published four contributions on the nature and cause of thunder.)

The first of the theories held that the lightning stroke creates a vacuum and that thunder is produced when the vacuum collapses. Another maintained that water drops in the path of the lightning flash are turned into steam and that the rapid expansion of the steam is accompanied by a loud report. A third proposal suggested that the electrical discharge decomposes water molecules by electrolysis and that the hydrogen and oxygen thereby produced subsequently recombine explosively. Finally, the simplest explanation ascribed thunder to the sudden heating of the air in the path of the lightning flash. Because air has electrical resistance, it is heated by the passage of a current just as a wire is; the expansion of the heated air was considered sufficient to explain the ensuing thunder.

We now know that the last explanation is the correct one. Each surge of current in the lightning flash heats the air in its path, creating a channel of gases at high temperature and pressure. The gases expand into the surrounding air as a shock wave, which after traveling a short distance decays into an acoustic wave.

It is interesting to note that the other theories were not entirely without foundation. For example, a region of diminished air pressure is briefly created in the aftermath of lightning; this partial vacuum is an effect of thunder, however, rather than a cause. Moreover, water drops are certainly evaporated in the lightning channel, and water molecules are decomposed. There is excellent evidence for the decomposition in the optical spectrum of lightning: one of the most prominent features of the spectrum is an emission line of hydrogen. Both of these phenomena, however, are merely collateral effects of lightning; they make no significant contribution to thunder.

The interest in the nature of thunder that culminated in the theories of the early 20th century declined soon thereafter, and the study of thunder was largely ignored for 50 years. Isolated experiments and observations were made during that period, but they tended more to confuse than to enhance understanding. For example, from measurements of the duration of thunder one can estimate the length of the lightning channel; in many cases such measurements were found to yield lengths much greater than that of the observed lightning flash and in some cases the lengths were greater than the height of the cloud itself. This discrepancy has only recently been explained.

Since 1960 scientific interest in thunder has revived and research into the nature and source of thunder has been renewed. The investigation is greatly assisted by a technology that was unavailable to earlier workers. The fundamental principles are no longer in question; the challenge to research today is to account for the detailed features of the thunder signature.

What is heard in a peal of thunder depends in large measure on the characteristics of the particular lightning flash that produced it. Both the temporal sequence of events in the discharge and their arrangement in space must be considered; these two complex factors determine not only the frequency and the amplitude of the radiated acoustic waves but also the order in which the waves

40

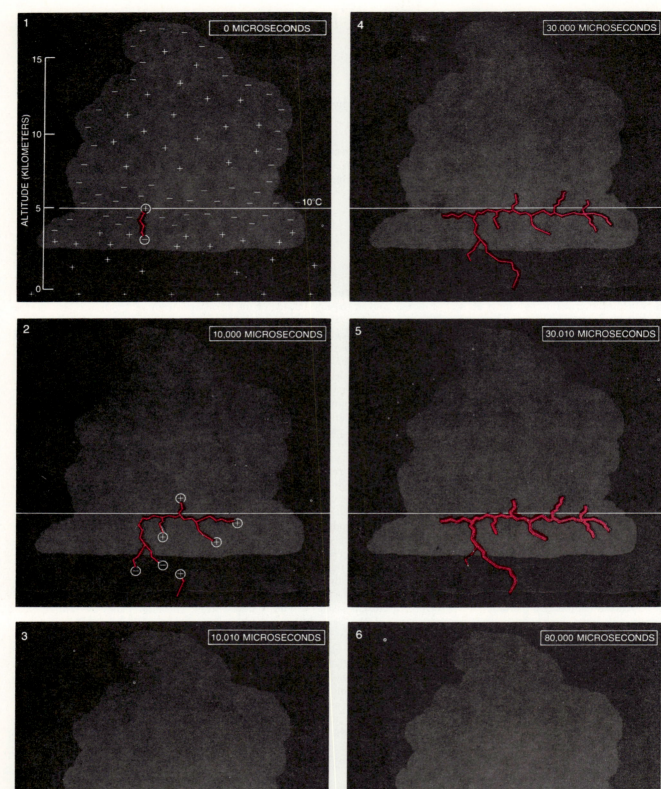

are received at a given observation site.

A lightning flash in most instances begins near the base of a cloud in a region of dense negative charge. This stratum is typically at an altitude of about five kilometers, where the temperature is approximately −10 degrees Celsius. That is the region of the cloud where water droplets freeze, a process that may be connected with the generation of electric charge.

The negatively charged zone of the cloud can be at a potential of as much as 300 million volts with respect to the ground, but even that enormous potential is insufficient to support a spontaneous arc across five kilometers of air. The main discharge can begin only after the channel has been traced by a preliminary low-current discharge called the stepped leader. The stepped leader begins to form when electrons emitted by droplets in the cloud are accelerated by the intense electric field; the electrons collide with molecules in the air, freeing many more electrons and leaving a conductive path of partially ionized air. Such a cascade of accelerated electrons typically progresses only 50 to 100 meters, but with each step a portion of the cloud's charge is transferred downward, and the next step can begin from the tip of the advancing leader.

The course of the stepped leader is highly irregular and in its progress toward the ground it forms numerous branches. Each step is accomplished in less than a microsecond, but there is a pause of about 50 microseconds between steps. As the leader approaches the ground the potential gradient—the voltage per meter—increases, and sparks are emitted from objects and structures on the ground, usually from the highest points first. When one of these sparks meets the downward-propagating leader, a conducting path between the cloud and the ground is completed. Since the potential difference across the path is a few hundred million volts a surge of current immediately follows; this large current is called the first return stroke.

The stepped leader may require 20 milliseconds to create the channel to the ground, but the return stroke is completed in a few tens of microseconds. In some cases that is the end of the lightning flash; more commonly, however, the leader-and-stroke process is repeated in the same channel at intervals of tens to hundreds of milliseconds. The subsequent leaders, called dart leaders, progress faster and more smoothly than the stepped leader because the electrical resistance of the path they follow is lower than that of the surrounding air. As the dart leader progresses toward the ground, intracloud processes extend the channel within the cloud so that additional areas are discharged. The subsequent return strokes, however, are usually less energetic than the first one is. A typical lightning flash has three or four leaders, each followed by a return stroke; one flash has been photographed that had 26 strokes.

Each surge of current in the flash, including the steps in the stepped leader, the dart leaders and the return strokes, heats the gases in the lightning channel and thereby generates an acoustic signal. The amplitude and duration of the signal produced by each current surge depend on the magnitude of the current. The complete thunder signal therefore reflects a complex sequence of events in the lightning flash. Ordinarily it is not possible to distinguish in a recorded thunder signature the acoustic pulses generated by individual leaders and strokes, but the sequence of strokes and the currents they carry nevertheless determine what sound is produced. In a few exceptional recordings the correlation of thunder pulses with lightning strokes is evident.

The spatial arrangement of the lightning flash has perhaps a greater influence on the resulting thunder than the temporal organization. The lightning channel is said to be tortuous: it consists of straight segments separated by sharp bends. The structural elements are classified in three size ranges. The large-

scale features of the channel, called the macrotortuous elements, are straight segments at least 100 meters long. Straight segments between five and 100 meters long are classified as mesotortuous, and those shorter than five meters are called microtortuous. The mesotortuous and macrotortuous elements are not, of course, actually straight lines; they are made up of many smaller segments and seem to be straight only when considered at the proper scale. Moreover, the classification of a feature as mesotortuous or microtortuous is not determined by a fixed rule but depends on the energy of the discharge. Five meters is an appropriate minimum length for mesotortuous features in strokes carrying large currents, but smaller elements can be considered mesotortuous when the current is comparatively small, as it is in stepped leaders and dart leaders.

The mesotortuous segments are the primary radiators of the acoustic pulses of thunder. The entire channel can be considered a "string of pearls," each pearl a mesotortuous segment radiating a series of pulses determined by the sequence of pulses in the lightning flash. The macrotortuous segments determine the spatial organization of the individual acoustic radiators and hence have a profound influence on what is heard by the observer.

A pulse of thunder begins in a channel of hot gases at high pressure. Spectroscopic evidence indicates that the temperature in the channel can reach 30,000 degrees C., and the pressure can exceed atmospheric pressure by from 10 to 100 atmospheres. Initially the high-pressure core expands as a shock wave. A shock wave is distinguished from an acoustic wave in that it compresses and heats the medium in which it propagates and thereby increases the speed of sound. Because the speed of sound increases as the temperature rises, the shock wave travels faster than sound does in the same medium. The extent of the compression and heating, and therefore the magnitude of the increase in velocity, depend on the amplitude of the wave. Behind the shock wave the air continues to move outward, and a region of low pressure forms.

The expanding shock wave dissipates its energy in performing work on the surrounding air. When all the energy imparted to the shock wave by the lightning stroke has been expended, the wave "relaxes" and the pressure in the vicinity of the channel returns to normal. (The core of the channel itself, however, re-

LIGHTNING CHANNEL develops from a region of concentrated negative charge near the base of a cloud, an area associated with the freezing of water droplets. The flash begins with the formation of a stepped leader (1), which moves downward in steps from 50 to 100 meters long and simultaneously extends streamers horizontally through the charged region. As the leader nears the ground, sparks propagate upward to meet it (2); when the pathway is completed, the large current of the first return stroke flows (3). Subsequent leaders, called dart leaders (4), progress much faster and extend the channel to other parts of the cloud; each is followed by a return stroke (5). Much of the length of the completed channel is horizontal and only a small portion of it is visible below the cloud (6). Each leader and stroke heats the gases in the lightning channel and contributes to the signal ultimately perceived as thunder.

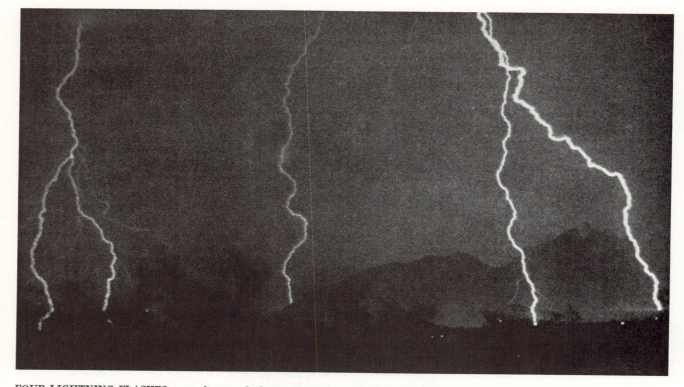

FOUR LIGHTNING FLASHES were photographed near Tucson, Ariz. The prominent kinks and bends are the large-scale, or macrotortuous, elements of the lightning channels; segments between major bends are on the order of 100 meters long. The macrotortuous features determine the overall pattern of claps and rumbles in thun-der. The four flashes were not simultaneous but were recorded in a single photograph by making a time exposure lasting for about two minutes. The photograph was made by Henry B. Garrett of Rice University. The forked lightning channel at left was produced by a flash consisting of multiple strokes that followed divergent paths.

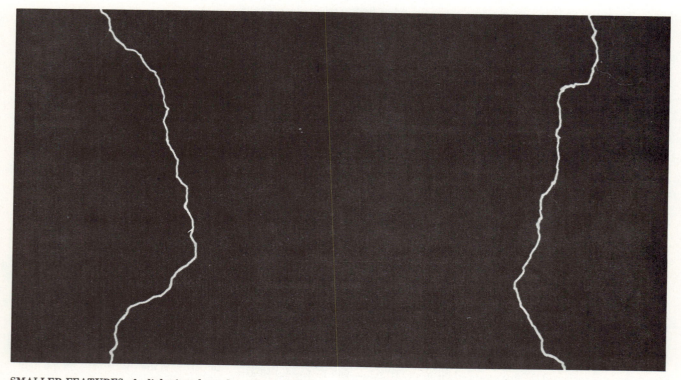

SMALLER FEATURES of a lightning channel are discernible in a photograph made with a telephoto lens. The area seen is the lower portion of the forked channel in the photograph at the top of the page. The many small straight segments are on the order of 10 me-ters long; elements of the channel in this size range are classified as mesotortuous and can be considered point sources of thunder. The microtortuous features (those shorter than about five meters) can rarely be resolved in photographs and have little effect on thunder.

mains a region of high temperature and low density.) From the ambient pressure and the energy per unit length of the lightning stroke one can calculate the radius at which the shock wave relaxes. The relaxation radius is in turn related to the wavelength of the thunder. Thus the wavelength, or pitch, of thunder is determined by the energy of the lightning stroke and the ambient air pressure in the region where the thunder is generated. The more powerful the stroke or the lower the air pressure, the lower the pitch of the resulting thunder. A typical value is 60 hertz [see illustration below].

The relaxation radius serves as the actual measure distinguishing microtortuous elements from mesotortuous ones. Features of the channel that are smaller than the relaxation radius are blurred in the expansion of the shock waves. They have little influence on the form of the resulting acoustic waves and therefore cannot be resolved in data derived from thunder. Microtortuous features are thus defined as those that are too small to be detected in the thunder signature.

The shock wave is not an efficient source of acoustic radiation: less than 1 percent of its energy is transmitted to the acoustic wave. The remaining 99 percent is dissipated in heating the air in the vicinity of the lightning channel. The total energy of the shock wave is very large, however, and even the small fraction of it converted into sound generates an acoustic wave of large amplitude. The result is one of the loudest sounds in nature.

Sound produced by the mesotortuous features of the channel is not radiated with equal power in all directions. The study of large sparks produced in the laboratory has shown that more than 80 percent of the acoustic energy is confined to a zone 30 degrees above and below a plane that bisects the spark perpendicularly. A microphone placed end on to the spark will receive much less sound than one placed broadside to it. This directed quality of the radiated sound is one of the most important determinants of what is heard by the listener on the ground.

From studies of photographs of lightning it has been found that the average change in direction or orientation between adjacent mesotortuous segments is about 16 degrees. Because this angle is substantially smaller than the zone 60 degrees wide into which each segment radiates most of its energy, a number of mesotortuous segments strung along a single macrotortuous segment will radiate most of their energy in roughly the same direction [see top illustration on following page]. It is this property that is responsible for the sudden claps and prolonged rumbles of thunder.

Acoustic pulses from all parts of a lightning stroke are, of course, emitted almost simultaneously. The entire acoustic output of a lightning flash is produced in less than a second, in the time required for the complete sequence of leaders and strokes. The thunder persists much longer than that because the lightning channel is long (at least five kilometers long and often considerably more) and signals from its nearest segments reach the listener sooner than those from the farthest extensions. The orientation of the channel, and in particular the orientation of the macrotortuous segments, determine the character of the sounds heard.

If a macrotortuous segment is oriented end on to the observer, the sound received will be of comparatively low amplitude, since most of the acoustic energy is radiated perpendicularly to the segment. Moreover, the wave fronts from each mesotortuous element will reach the listener in succession, beginning with signals emitted by the nearest part of the segment. The result is a protracted rumble or roll of thunder.

If the macrotortuous segment is broadside to the listener, a much larger portion of the radiated energy will be received. Equally important, the wave fronts from all the mesotortuous elements will arrive almost simultaneously. As a result a brief but intense clap of thunder will be heard [see bottom illustration on following page].

Thunder produced by a single flash of lightning is commonly perceived as a combination of claps and rumbles because various elements of the lightning channel are oriented differently with respect to the listener. Thunder from the same lightning flash will be perceived differently at different locations, since each location has a unique position and orientation with respect to the lightning channel.

Between the lightning channel and the listener the thunder signal is altered by the medium in which it travels. The atmosphere attenuates, scatters and refracts the signals; in addition they are subject to nonlinear propagation effects

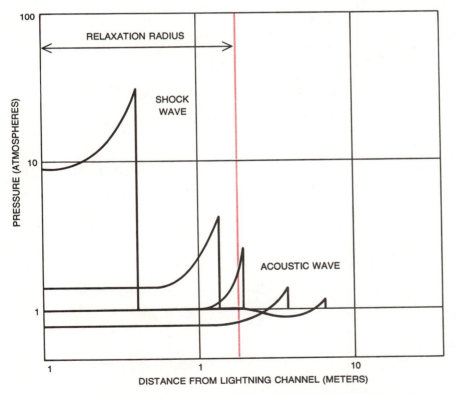

EXPANDING CHANNEL of hot gases produced by a lightning stroke propagates as a shock wave and then as an acoustic pulse of thunder. Because the initial shock wave compresses and heats the air, it quickly dissipates its energy. A few meters from the lightning channel it "relaxes" to yield an acoustic wave of lower amplitude, and the pressure in the region behind the wave is briefly reduced. Only about 1 percent of the energy of the shock wave is transferred to the acoustic wave; the rest is expended in heating the air near the channel.

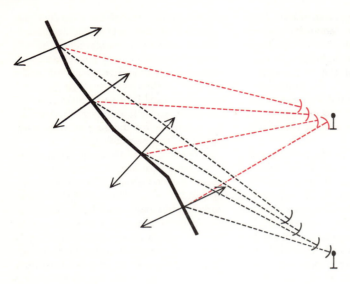

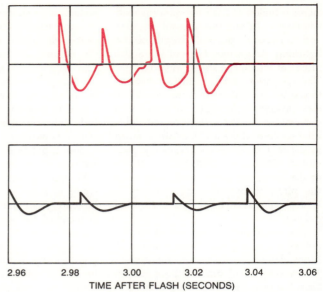

TIME AFTER FLASH (SECONDS)

MESOTORTUOUS SEGMENTS of the lightning channel are the primary radiators of thunder. Each of the four segments shown can be considered an independent point source of sound. Acoustic pulses are emitted from the segments almost simultaneously, but the pattern of sounds perceived on the ground depends on how the channel is oriented with respect to the observer. Sounds emitted perpendicularly to a segment (*color*) are more powerful than sounds emitted parallel or nearly parallel to it (*black*). Since each segment differs in orientation from the adjoining segments by only a small angle, signals from several segments reach an observer whose position is perpendicular to the channel almost simultaneously; the result is a brief but loud clap of thunder. An observer looking up the length of the channel, on the other hand, receives wave fronts at greater intervals and hears a prolonged rumble.

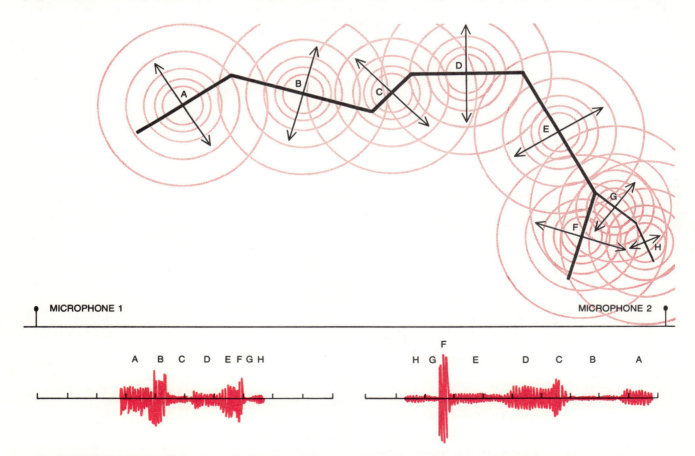

MACROTORTUOUS ELEMENTS of the lightning channel determine the overall pattern of claps and rumbles in the thunder signature. The amplitude and duration of the signal produced by each segment are determined by the orientation of the segment and its distance from the observer. Microphones placed at different positions will record different signatures from the same lightning channel. For this hypothetical lightning flash the segments have been labeled along with the sounds they would produce at two locations.

and, once they reach the ground, to reflection.

"Nonlinear propagation" refers to processes that affect one part of a wave more than others or some frequencies of sound more than others. Nonlinear propagation therefore alters the waveform of the signal (the shape of the individual waves) or its spectrum. One such effect tends to lengthen each pulse as it propagates; sound waves of large amplitude are the most severely affected, so that the process is most significant close to the lightning channel.

The attenuation of thunder in the atmosphere is caused by two independent processes. "Classical" attenuation results from the viscosity of air, that is, from the fact that air is not a perfectly elastic medium. Classical attenuation is well understood, and its magnitude can be predicted. "Molecular" attenuation results from a complex interaction of sound waves with water molecules and oxygen molecules in which acoustic energy excites internal vibrations of the molecules. It can be evaluated only if the temperature and humidity are known for all points along the path traveled by the thunder signal. Molecular attenuation is usually the more important of the two effects. Attenuation increases as the square of the frequency of the signal; thus a 20-hertz wave will travel four times as far as a 40-hertz wave before it is attenuated to the same degree.

The scattering of the thunder signal is even more difficult to predict than its attenuation. The principal agents of scattering are turbulent eddies in the atmosphere, which range in size from microscopic disturbances a few microns in diameter to the thunderstorm cell itself. For the scattering of thunder the most significant eddies are those that are approximately the size of the wavelength of the thunder (less than 50 meters). Again, the losses are severer for higher frequencies.

Because of scattering and attenuation, when a thunder signal travels several kilometers through a turbulent medium, only the lowest-frequency components of the original spectrum survive without major modification. As a result a lightning flash of low energy, which produces little low-frequency sound, will not give rise to audible thunder except at close range.

The refraction of thunder is a large-scale phenomenon caused by variations in the speed of sound in the atmosphere; a refracted "ray" of sound is bent away from the straight-line path between the source and the observer. The laws governing refraction are well understood, and given sufficient information on atmospheric conditions, the curved path followed by the acoustic ray can be calculated. The most important variables are temperature and wind.

In the lower atmosphere temperature ordinarily decreases with altitude, typically at a rate of about 6.5 degrees C. per kilometer. Below a thundercloud the temperature gradient is generally steeper, reaching a maximum of 9.8 degrees per kilometer. Because sound travels faster in warm air than it does in cool air, the temperature gradient tends to curve the acoustic rays upward [see illustration below]. For this reason thunder generated by the lowest parts of a lightning channel cannot be heard at a distance, and there is a distance beyond which thunder cannot be heard at all, since all the sound passes over the head of the observer.

Wind has two effects on the propagation of sound waves. First, the actual velocity of a wave front is the sum of the speed of sound in air and the wind velocity. Sound therefore propagates downwind faster than it does upwind, and in a precise analysis of thunder signals this difference must be taken into account. Second, wind shear, the variation of wind velocity with altitude, imparts a further bend to the acoustic rays. Wind velocity generally increases with altitude. In the upwind direction the effects of wind shear add to those of temperature refraction; in the downwind direction they subtract from them.

In combination with attenuation and scattering, the temperature gradient and wind shear impose an ultimate limit on the range over which thunder can be heard. The maximum range can be as little as 10 kilometers, but it is sometimes much greater, depending mainly on the altitude of the lightning channel and the wind velocity.

Reflection adds a final modification to the thunder signal before it reaches the listener. For low-frequency sound the amplitude of a reflected wave is roughly proportional to the angular size of the

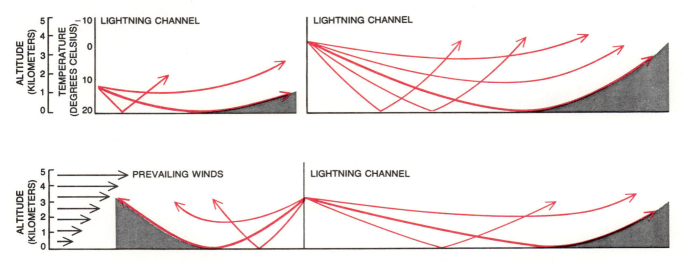

REFLECTION AND REFRACTION alter the thunder signal and limit its range. The temperature gradient in the atmosphere bends sound waves upward, so that thunder emitted from the lower portions of the lightning channel can be heard only nearby (*top left*); there is a maximum range beyond which thunder cannot be heard (*top right*). The distribution of sound produced by temperature refraction is altered by the wind, which usually has progressively higher velocity at higher altitude. Upwind of the channel the refraction is increased; downwind it is diminished (*bottom*). Low-frequency sound waves striking the ground are efficiently reflected.

reflecting surface as it is viewed by the observer. In most cases there is only one reflecting surface large enough to produce amplitudes comparable to those of signals received directly: the ground. Since flat ground completely fills the field of view when one is looking downward, the amplitude of the reflected wave will be equal to that of the directly received wave if no sound is absorbed at the surface, which for low frequencies is probably a correct assumption. The perceived sound is thus the sum of the direct and the reflected waves.

Depending on one's height above the ground and the angle of the incident signal, the direct and the reflected waves will add constructively for some wavelengths and destructively for others. In recording thunder signatures we avoid destructive interference by placing our microphones at a height equal to a fraction of the shortest wavelength to be measured.

The thunder perceived at any given position is unique to that point of observation. A microphone placed a few meters above an identical microphone at ground level will detect a slightly different signal. If the two microphones are both near the ground but are separated by a horizontal distance of 20 to 30 meters, the recorded thunder signatures will be similar, but minor differences will be distinguishable because the two positions give a significantly different perspective on the mesotortuous elements of the lightning channel. If the microphones are placed 100 meters apart, few details of the recorded signatures will correspond, but the underlying pattern of claps and rumbles will be preserved; the spacing is now comparable to the size of the macrotortuous elements. Microphones separated by more than a few kilometers will record thunder signatures that may have only one or two features in common.

A thunder signature contains a great deal of information about the lightning channel that produced it and about the atmosphere between the lightning and the observation site. The signature is a complex waveform, but it is possible to analyze it and recover much of the information. When that is done, one can reconstruct the lightning channel, a particularly valuable capability when the channel runs inside a cloud and photography and other optical methods of study are not possible.

The thunder signature is recorded by

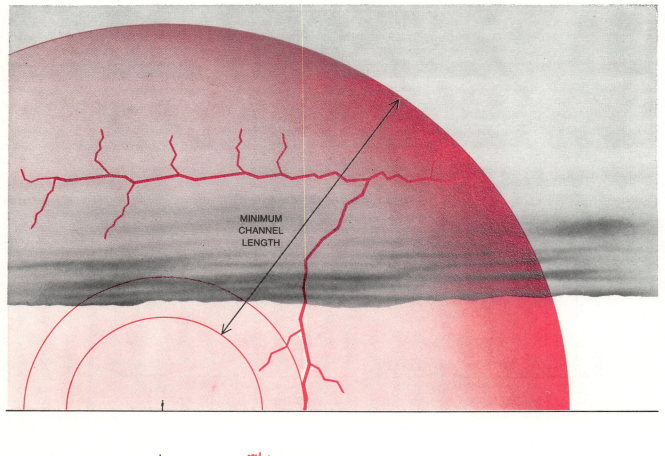

MINIMUM
CHANNEL
LENGTH

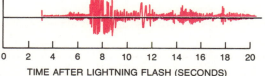

TIME AFTER LIGHTNING FLASH (SECONDS)

AMATEUR OBSERVATIONS of thunder require only a wristwatch. By measuring the time elapsed from the lightning flash to the first thunder heard, the loudest clap of thunder and the last rumble, one can determine the distance to the nearest branch of the lightning channel, the main channel and the farthest feature. The approximate distance in kilometers is given by the time in seconds divided by 3. By the same method the minimum length of the channel can be calculated from the total duration of the thunder signal.

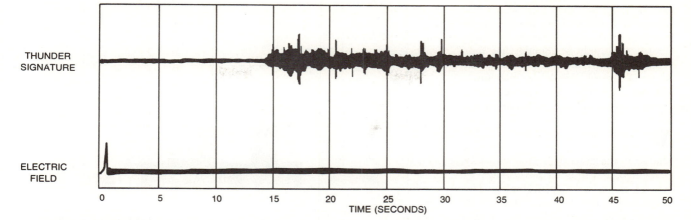

THUNDER
SIGNATURE

ELECTRIC
FIELD

TIME (SECONDS)

THUNDER SIGNATURE recorded by one microphone in an array of several microphones contains detailed information about the lightning channel that produced it. The time of the lightning flash is recorded by the momentary change in the electric field at the beginning of the graph. From the time of arrival of each feature of the thunder signature the distance to the corresponding segment of the channel can be calculated; with information from additional microphones the direction can also be determined and hence the location of the channel. Corrections must also be made for effects introduced by the wind and by temperature variations in the atmosphere.

an array of microphones on flat ground. For the purposes of analysis the thunder signal is broken up into short sections, each one-fourth of a second to half a second long. The microphones measure the direction of arrival of each section, and in addition the time of the lightning flash and the time of arrival of each section of the signature are known. With the aid of a digital computer and a mathematical model of the atmosphere it is then possible to determine the origin of each section, that is, to deduce its position at the time of the lightning flash. In this way the entire channel can be reconstructed in three-dimensional form. The technique is sensitive enough to locate the main channel and some of the larger branches, but most small branches are unrecoverable because the thunder produced by them has too small an amplitude.

By studying the acoustic record of an entire thunderstorm we have been able to compile a graphic history of the large-scale electrical discharges during the lifetime of a storm. From the same data we have derived information on the process by which the cloud acquires elec-

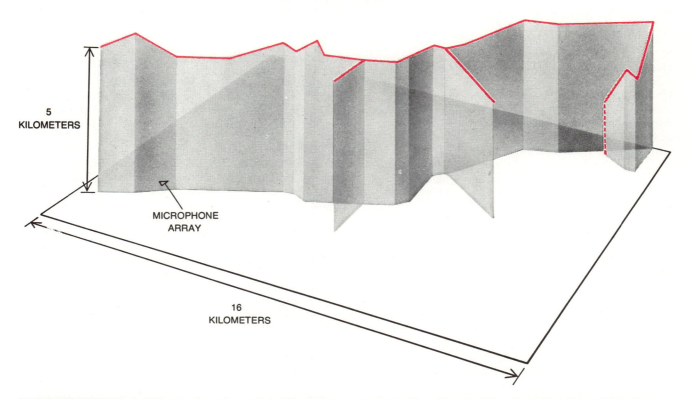

5
KILOMETERS

MICROPHONE
ARRAY

16
KILOMETERS

RECONSTRUCTION of a lightning channel was derived from the thunder signature at the top of the page. Most of the channel is horizontal and is hidden from visual observation by cloud. The macrotortuous segments of the main channel and the larger branches are recorded, but smaller features cannot be resolved and small branches cannot be reconstructed because the thunder they produce is too feeble. Thunder from lowest portion of the channel (*broken line*) could not be detected at the microphone array because of refraction.

tric charge, the volume in which the charge is stored and the time required to replenish the charge after a lightning stroke. The study of thunder has thus made possible tentative generalizations about electrical activity in the atmosphere and has led to some surprising discoveries about lightning inside clouds.

We have found that intracloud lightning (lightning that does not discharge to the ground) is predominantly horizontal, and so is the intracloud portion of cloud-to-ground lightning. The horizontal lightning channels tend to be aligned so that most of them are roughly parallel.

Our results also show that the negative charge center in the lower part of the cloud is commonly disk-shaped, about two kilometers thick and 10 kilometers in diameter. The positive charge, on the other hand, appears to be dispersed throughout the upper part of the cloud. The negatively charged region is consistently located near the -10 degree isotherm, an indication that the development of the charge is related to the freezing of droplets or to the coexistence of ice and droplets in the same part of the cloud. The development of charge also seems to be correlated with regions of inflow or updraft, which contain small droplets, rather than with areas where raindrops are found.

Lightning early in the life of a storm is confined to the lower, negatively charged region; the upper zone becomes active late in the cycle. Successive lightning flashes draw charge from different volumes of the cloud, but the channels frequently intersect at some point. It is as if one flash took up where the previous one left off. Moreover, lightning in one part of the cloud frequently triggers a discharge in another region. Certain processes important in the physics of clouds, such as the growth of cloud particles, appear to be strongly influenced by the electric field and are correlated with lightning activity.

Finally, the average length of the lightning channel seems to vary with the type of storm. Small, local storms have relatively short lightning discharges (typically about five kilometers long), and all the lightning channels in the storm fall within a narrow range of lengths. In storms associated with large frontal systems, on the other hand, the lightning channels have a broad distribution of lengths and a large mean value (about 15 kilometers). In the larger storms much of the channel length is horizontal.

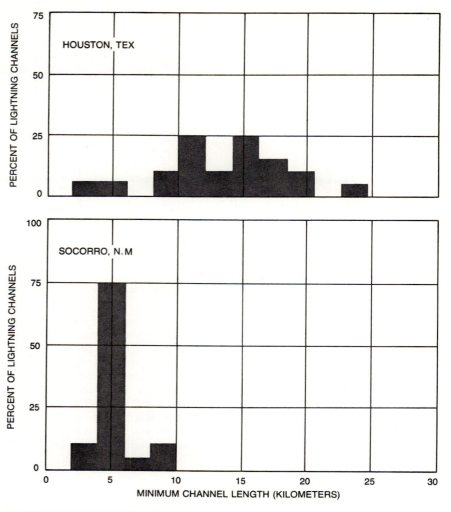

LENGTH OF LIGHTNING CHANNELS varies from storm to storm. A thunderstorm near Houston, Tex., associated with a frontal system, had a broad distribution of lengths, including many quite long channels. In a local storm at Socorro, N.M., however, there were no very long channels and the great majority were clustered around a single value. The lengths measured were the minimum possible lengths, as determined by duration of the thunder signal.

The complete reconstruction of a lightning channel requires arrays of microphones sensitive to low-frequency sound, equipment capable of recording the sound and a computer. It is possible for the amateur observer, however, to recover some of the information from the thunder signature with apparatus no more elaborate than a wristwatch. You can obtain the approximate distance to the source of a feature in the thunder signature by multiplying the time elapsed between the lightning flash and the arrival of the acoustic signal by the speed of sound. (The approximate distance in kilometers is given by the time in seconds divided by 3, the distance in miles by the time in seconds divided by 5.) Similarly, you can estimate the minimum length of the channel from the total duration of the thunder, again multiply-

ing the elapsed time by the speed of sound [see illustration on page 46].

With careful observation you may be able to distinguish some of the individual events that make up a cloud-to-ground lightning flash: the creation of the tortuous, branched channel forged by the stepped leader, the brightening of the channel with the first return stroke and a flickering produced by multiple return strokes. Occasionally a lightning flash forks when a dart leader deviates from the path of the original channel.

During a thunderstorm at night you might also want to try photographing lightning. Set up the camera on a tripod and point it toward the most active region of the storm. Close the iris to the smallest aperture possible (that is, set the lens to the largest f number), focus at infinity and make time exposures of from 20 to 30 seconds. Typically one in every three or four frames will contain a lightning photograph, although the number depends on the activity of the storm.

More information about the lightning channel can be obtained through careful attention to the thunder signature. Measure the delay between the flash and the first thunder heard, the loudest clap and the final rumble. From these times you can estimate respectively the distance to the branch nearest you, to the main channel and to the farthest branch. Also note the total duration of the thunder in order to calculate the minimum length of the channel.

A discharge that has struck the ground nearby gives rise to a loud crack, sometimes preceded by a brief rumble or a ripping noise that probably originates in a small branch extending from the main channel toward the observer. When a nearby flash is made up of several strokes, it is sometimes possible to distinguish the acoustic pulses produced by each stroke. The sound is somewhat like a short burst of machine-gun fire.

When thunder powerful enough to shake windows is heard as a boom rather than as a clap or a crash, it is usually the product of an energetic flash but a distant or high-altitude one. At a distance of a few kilometers the high frequencies are attenuated with respect to the low frequencies, and the resulting thunder is felt as much as it is heard. In some cases the pitch of the thunder grows progressively lower as sounds arrive from higher or more distant portions of the channel.

Finally, a somewhat rare thunder signal is a ripping noise that can be imagined as the tearing of some cosmic cloth. It is usually attributed to a stepped leader that fails to reach the ground. When the leader does complete a path to the ground, the sound it generates is overwhelmed by that of the subsequent return stroke.

Obviously caution is in order when you are observing a thunderstorm. Do not expose yourself to the risk of personal participation in a lightning discharge by standing in open country or near trees, power lines, fences or other objects likely to be struck. The safest observation post is a closed space, such as a building or an automobile, provided that you avoid touching exterior walls and conducting surfaces.

6   Why the Sea Is Salt

by Ferren MacIntyre
November 1970

The sea contains more than 70 elements in addition to sodium and chlorine. The global cycles that remove and replenish them involve rainfall, volcanoes and the spreading of the ocean floor.

According to an old Norse folktale the sea is salt because somewhere at the bottom of the ocean a magic salt mill is steadily grinding away. The tale is perfectly true. Only the details need to be worked out. The "mill," as it is visualized in current geophysical theory, is the "mid-ocean" rift that meanders for 40,000 miles through all the major ocean basins. Fresh basalt flows up into the rift from the earth's plastic mantle in regions where the sea floor is spreading apart at the rate of several centimeters per year. Accompanying this mantle rock is "juvenile" water—water never before in the liquid phase—containing in solution many of the components of seawater, including chlorine, bromine, iodine, carbon, boron, nitrogen and various trace elements. Additional juvenile water, equally salty but of somewhat different composition, is released by volcanoes that rim certain continental margins, such as those bordering the Pacific, where the sea floor seems to be disappearing into deep trenches [*see illustration on these two pages*].

The elements most abundant in juvenile water are precisely those that cannot be accounted for if the solids dissolved in the sea were simply those provided by the weathering of rocks on the earth's surface. The "missing" elements, such as chlorine, bromine and iodine, were once called "excess volatiles" and were attributed solely to volcanic emanations. It is now recognized that juvenile water may have nearly the same chlorinity as seawater but is much more acid due to the presence of one hydrogen ion (H^+) for every chloride ion (Cl^-). In due course, as I shall explain later, the hydrogen ions are removed and replaced by sodium ions (Na^+), yielding the concentration of ordinary salt (NaCl) that constitutes 90-odd percent of all the "salt" in the sea.

The chemistry of the sea is largely the chemistry of obscure reactions at extreme dilution in a strong salt solution, where all the classical chemist's "distilled water" theories and procedures break down. The father of oceanographic chemistry was Robert Boyle, who demonstrated in the 1670's that fresh waters on the way to the sea carry small amounts of salt with them. He also made the first attempt to quantify saltiness by drying seawater and weighing the residue, but his results were erratic because some of the constituents of sea salt are volatile. Boyle found that a better method was simply to measure the specific gravity of seawater and from this estimate the amount of salt present. Since the distribution of density in the sea is important to oceanographers, the same calculation is routinely performed today in reverse: the salinity is deduced by measuring the electrical conductivity of a sample of seawater, and from this and the original temperature of the sample one can compute the density of the seawater at the point the sample was taken.

In 1715 Edmund Halley suggested that the age of the ocean and thus of the world might be estimated from the rate of salt transport by rivers. When this proposal was finally acted on by John Joly in 1899, it gave an age of some 90 million years. The quantity that Joly measured (total amount of x in ocean divided by annual river input of x) is now recognized as the "residence time" of the constituent x, which is an index of an element's relative chemical activity in the ocean. Joly's value is about right for the residence time of sodium; for a more reactive element (in the ocean environment) such as aluminum the residence time is as brief as 100 years.

Not quite 200 years ago Antoine Laurent Lavoisier conducted the first analysis of seawater by evaporating it slowly and obtaining a series of compounds by fractional crystallization. The first compound to settle out is calcium carbonate ($CaCO_3$), followed by gypsum ($CaSO_4 \cdot 2H_2O$), common salt (NaCl), Glauber's salt ($Na_2SO_4 \cdot 10H_2O$), Epsom salts ($MgSO_4 \cdot 7H_2O$) and finally the chlorides of calcium ($CaCl_2$) and mag-

MAGIC SALT MILL at the bottom of the sea, imagined in the old Norse folktale, turns out to be not so fanciful after all. The modern explanation of why the sea is salt invokes the concept of the "mid-ocean" rift and sea-floor spreading, as depicted here in cross section. The rift is a weak point be-

nesium (MgCl$_2$). Lavoisier noted that slight changes in experimental conditions gave rise to large shifts in the relative amounts of the various salts crystallized. (In fact, some 54 salts, double salts and hydrated salts can be obtained by evaporating seawater.) To get reproducible results for even the total weight of salt one must remove all organic matter, convert bromides and iodides to chlorides, and carbonates to oxides, before evaporating. The resulting weight, in grams of salt per kilogram of seawater, is the salinity, S‰. (The symbol ‰ is read "per mil.")

In actual practice the total weight of salt in seawater is nowadays never determined. Instead the amount of chloride ion is carefully measured and a total for all other ions is computed by applying the "constancy of relative proportions." This concept dates back to the middle of the 19th century, when John Murray eliminated confusion about the multiplicity of salts by observing that individual ions are the important thing to talk about when analyzing seawater. Independently A. M. Marcet concluded from many measurements that various ions in the world ocean were present in nearly constant

proportions, and that only the absolute amount of salt was variable. This constancy of relative proportions was confirmed by Johann Forchhammer and again more thoroughly by Wilhelm Dittmar's analysis of 77 samples of seawater collected by H.M.S. *Challenger* on the first worldwide oceanographic cruise. These 77 samples are probably the last ever analyzed for all the major constituents. Their average salinity was close to 35‰, with a normal variation of only ±2‰.

In the 86 years since Dittmar reported eight elements, 65 more elements have been detected in seawater. It was recognized more than a century ago that elements present in minute amounts in seawater might be concentrated by sea organisms and thereby raised to the threshold of detectability. Iodine, for example, was discovered in algae 14 years before it was found in seawater. Subsequently barium, cobalt, copper, lead, nickel, silver and zinc were all detected first in sea organisms. More recently the isotope silicon 32, apparently produced by the cosmic ray bombardment of argon, has been discovered in marine sponges.

There are also inorganic processes in

the ocean that concentrate trace elements. Manganese nodules (of which more below) are able to concentrate elements such as thallium and platinum to detectable levels. The cosmic ray isotope beryllium 10 was recently discovered in a marine clay that concentrates beryllium. In all, 73 elements (including 13 of the rare-earth group) apart from hydrogen and oxygen have now been detected directly in seawater [see illustration on page 53].

It is only in the past 40 years that geochemists have become interested in the chemical processes of the sea for what they can tell us about the history of the earth. Conversely, only as geophysicists have pieced together a comprehensive picture of the earth's history has it been possible to bring order into marine chemistry.

The earth's present atmosphere and ocean are not primordial but have been liberated from chemical and mechanical entrapment in solid rock. Perhaps four billion years ago, or a little less, there was (according to many geophysicists) a "grand catastrophe" in which the earth's core, mantle, crust, ocean and atmo-

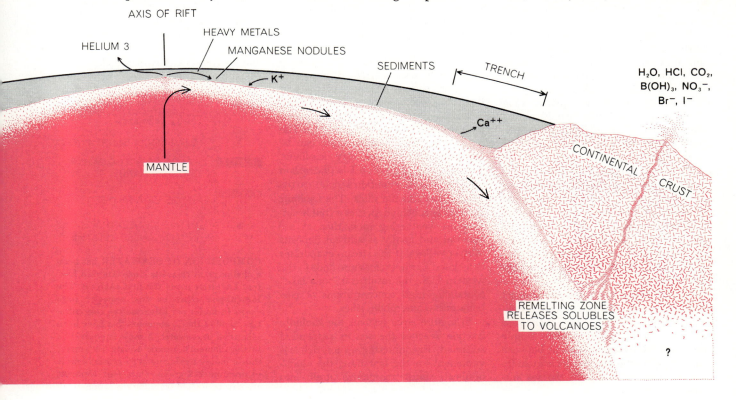

tween rigid plates, or segments, in the earth's crust. Although the driving mechanism is not yet understood, the plates move apart a few centimeters a year as fresh basalt from the plastic mantle flows up between them. The new basalt releases "juvenile" water (water never before in liquid form) and a variety of elements, including heavy metals that become incorporated in manganese nodules and the rare isotope helium 3, which escapes finally into space. At the

continental margin (*right*) the lithospheric plate is subducted, forming a trench and carrying accumulated sediments with it. (The plate apparently thickens en route as plastic basalt "freezes" to its underside.) As it descends the plate remelts and releases soluble elements and ions that are ejected into the atmosphere by volcanoes. They maintain the saltiness of the sea and together with weathered crustal rock, such as granite, provide the stuff of sediments.

sphere were differentiated from an original homogeneous accumulation of material. Estimates of water released during the catastrophe range from a third to 90 percent of the present volume of the ocean. The catastrophe is not finished even yet, since differentiation of the mantle continues in regions of volcanic activity. Most of the exhalations of volcanoes and hot springs are simply recycled ground water, but if only half of 1 percent of the water released is juvenile, the present production rate is sufficient to have filled the entire ocean in four billion years.

There is evidence that the salinity of the ocean has not changed greatly since the ocean was formed; in any event the salinity has been nearly constant for the past 200 million years (5 percent of geologic time). The composition of ancient sediments suggests that the ratio of sodium to potassium in seawater has risen from about 1 : 1 to its present value of about 28 : 1. Over the same period the ratio of magnesium to calcium has risen from roughly 1 : 1 to 3 : 1 as organisms removed calcium by building shells of calcium carbonate. It is significant, however, that the total amount of each pair of ions varied much less than the relative amounts.

If we look at rain as it reaches the sea in rivers, we find a distinctly nonmarine mix to its ions. If we catch it even earlier as it tumbles down young mountains, the differences are even more pronounced. This continual input of water of nonmarine composition would eventually overwhelm the original composition of the ocean unless there were corrective reactions at work.

The overall geochemical cycle that keeps the marine ions closely in balance involves a complex interchange of material over decades, centuries and millenniums among the atmosphere, the ocean, the rivers, the crustal rocks, the oceanic sediments and ultimately the mantle [see "a" in illustration on page 54]. Because this overall picture is too general to be of much use, we abstract bits from it and call them thalassochemical models (thalassa is the Greek word for "sea"). One model involves simply the cyclic exchange of sea salt between the rivers and the sea; the cycle includes the transport of salt from the sea surface into the atmosphere, where salt particles act as condensation nuclei on which raindrops grow [see "b" in illustration on page 54]. This process accounts for more than 90 percent of the chloride and about 50 percent of the sodium carried to the sea by rivers.

Another useful abstraction is the "steady state" thalassochemical model. If the ocean composition does not change with time, it must be rigorously true that whatever is added by the rivers must be precipitated in marine sediments [see "c" in illustration on page 54]. Oceanic residence times computed from sedimentation rates, particularly for reactive trace metals, agree well with the input rates from rivers. Unfortunately residence times do not reveal the mechanism by which an element is removed from seawater. For residence times greater than a million years it is often helpful to invoke the "equilibrium" model, which deals only with the rate of exchange between the ocean and its sediments [see "d" in illustration on page 54].

To understand how the earth maintains its geochemical poise over a billion-year time scale we must return to the circle of arrows—the weathering and "unweathering" processes—of the geochemical cycle. This circle starts with primordial igneous rock, squeezed from the mantle. Ignoring relatively minor heavy metals such as iron, we can assume that the rock consists of aluminum, silicon and oxygen combined with the alkali metals: potassium, sodium and calcium. The resulting minerals are feldspars (for example $KAlSi_3O_8$). Rainwater picks up carbon dioxide from the air and falls on the feldspar. The reaction of water, carbon dioxide and feldspar typically yields a solution of alkali ions and bicarbonate ions (HCO_3^-) in which is suspended hydrated silica (SiO_2). The residual detrital aluminosilicate can be approximated by the clay kaolinite: $Al_2Si_2O_5(OH)_4$ [see Step 1 in illustration on page 55]. A mountain stream carries off the ions and the silica. The kaolinite fraction lags behind, first as a friable surface on weathering rock, then as soil material and finally as alluvial deposits in river valleys. If the stream evaporates in a closed basin, such as one finds in the western U.S., the result is a "soda lake" containing high concentrations of carbonates and amorphous silica.

In mature river systems the kaolinite fraction reaches the sea as suspended sediment. Encountering an ion-rich environment for the first time, the aluminosilicate must reorganize itself into new minerals. One such mineral, which seems to be forming in the ocean today, is the potassium-containing clay illite [see Step 2 in illustration on page 55]. These "clay cation" reactions may take decades or centuries. They are poorly understood because graduate students

who study them invariably leave before the reactions are complete. The net effect of such reactions is to tie up and remove some of the potassium and bicarbonate ions, along with aluminum, silicon and oxygen.

A biologically important reaction, usually confined to shallow water, allows marine organisms to build shells of calcium carbonate, which precipitates when calcium (Ca^{++}) and bicarbonate ions react. If dilute hydrochloric acid is present (it is released by volcanoes), it reacts even more rapidly with bicarbonate, forming water and carbon dioxide and leaving free the chloride ion. When marine organisms die and sink to about 4,000 meters, they cross the "lysocline," below which calcium carbonate redissolves because of the high pressure. We have now traced the three metallic ions removed from igneous rock to three separate niches in the ocean. Sodium remains dissolved, potassium precipitates in clays on the deep-sea floor and calcium precipitates in shallow water as biogenic limestone: coral reefs and calcareous oozes.

Ages pass and the geochemical cycle rolls on, converting ocean-bottom clay into hard rock such as granite. When old sea floor finally reaches a region of high pressure and temperature under a continental block, it still contains some free ions that can react with the clay to reconstitute hard rock. A score of reaction

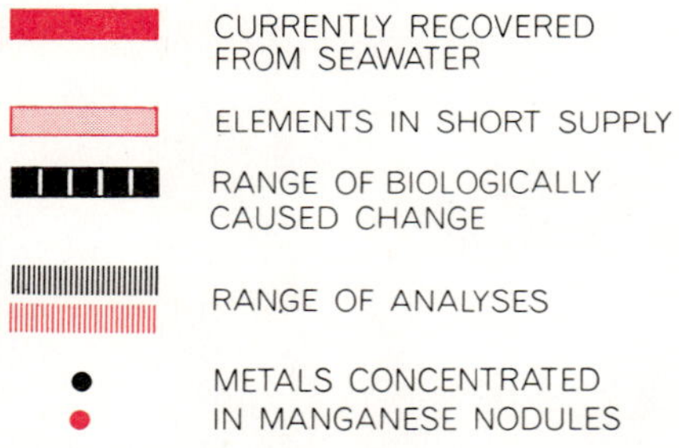

� (red filled bar)	CURRENTLY RECOVERED FROM SEAWATER
▭ (light red outline bar)	ELEMENTS IN SHORT SUPPLY
▐▐▐▐ (black striped bar)	RANGE OF BIOLOGICALLY CAUSED CHANGE
▐▐▐▐▐ (gray/red striped bar)	RANGE OF ANALYSES
● ● (black dot / red dot)	METALS CONCENTRATED IN MANGANESE NODULES

COMPOSITION OF SEAWATER has been a challenge to chemists since Antoine Laurent Lavoisier made the first analyses. The logarithmic chart on the opposite page shows in moles per kilogram the concentration of 40 of the 73 elements that have been identified in seawater. A mole is equivalent to the element's atomic weight in grams; thus a mole of chlorine is 35 grams, a mole of uranium 238 grams. Only four elements are now recovered from the sea commercially: chlorine, sodium, magnesium and bromine. Recovery of other scarce elements is not promising unless biological concentrating techniques can be developed. Manganese nodules are a potential source of scarce metals but gathering them from the deep-sea floor may not be profitable in this century.

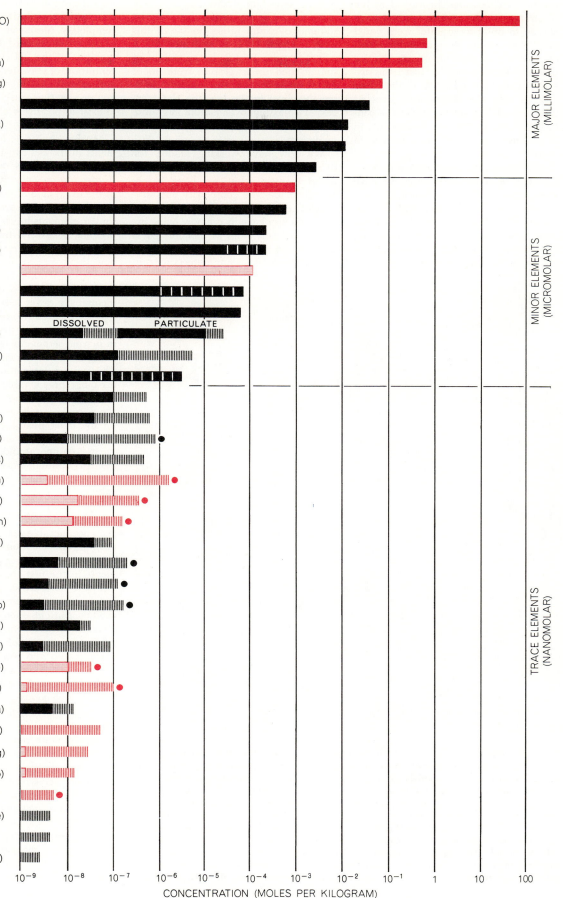

schemes are possible. In Step 3 in the illustration on the opposite page I have chosen to build a "granite" from equal parts of potassium feldspar, sodium feldspar, potassium mica and quartz. (Notice that calcium is missing because it has dissolved from the sediments during their descent into the deep-ocean trenches that carry the sediments under the continental blocks.) The reaction written in Step 3 uses up all the silica formed in Step 1.

The goal of this geochemical exercise has now been reached. First, we have shown that of all the substances that enter the ocean, only sodium and chlorine remain abundantly in solution. Of the other elements, the amount remaining in solution is less than a hundredth of the amount delivered to the ocean and

precipitated from it. Second, we have made a start at explaining the observed sodium-potassium ratios: in basalt this ratio is about 1 : 1, in seawater 28 : 1 and in granite 1 : 1.2. If the weight of sodium tied up in granite were about 140 times as great as the weight of sodium dissolved in the sea, the slight excess of potassium over sodium in granite would explain the sea's deficiency in potassium.

We now have working models for thinking about the circulation of the major elements, but we have barely scratched the true complexity and subtlety of seawater. The sources and sinks of the minor elements are now being explored. In many cases we can only guess at what the natural marine form of an element is because our detection tech-

niques either convert all forms to a common form for analysis or miss some forms completely. Moreover, certain ions seem to behave capriciously in the ocean. For example, at the pH (hydrogen-ion concentration) of seawater, vanadium should appear as $VO_2(OH)_3{}^{--}$, an ion with a double negative charge; instead it seems to exist in positively charged form, perhaps as $VO_2{}^+$.

Much of what is known about elements in the sea can be summarized in an oceanographer's periodic table [see illustration on page 56]. The usefulness of the usual kind of periodic table to the chemist is that it arranges chemically similar elements in vertical columns and presents behavioral trends in horizontal rows. The oceanographer's table shows how these regularities are disrupted in the ocean environment.

First of all, more than a dozen elements have never been detected in seawater, although two of them (palladium and iridium) exist in parts per billion in marine sediments and another (platinum) is present in manganese nodules. The second interesting feature of the oceanographer's table is the tendency for the "upper" and "outer" elements, those in the raised wings, so to speak, to be the most plentiful in the sea. The "upper" tendency simply reflects the greater cosmic abundance of light elements. (Lithium, beryllium and boron, however, are fairly scarce even cosmically.)

The "outer" trend can be explained in quantum-mechanical terms by the presence or absence of electrons in d orbitals, the electron shells principally involved in forming complexes. Elements in the first three columns at the left have no d orbitals; those in the last four columns at the right have full d orbitals. Both characteristics favor weak chemical bonds, with the result that these two groups of elements tend to ionize readily and remain in solution, either by themselves or in simple combination with oxygen and hydrogen. In contrast, the elements in the center of the table with partially filled d orbitals form strong chemical bonds and compounds that precipitate readily; thus they can exist only at low concentration in solution. For silver and the surrounding group of metals the most stable complexes are formed with the most abundant seawater ion: chloride. Most of the other elements that are hungry for d electrons form their complexes with oxygen, or oxygen plus some protons (hydrogen nuclei).

Ordinarily the oxidation state of metals

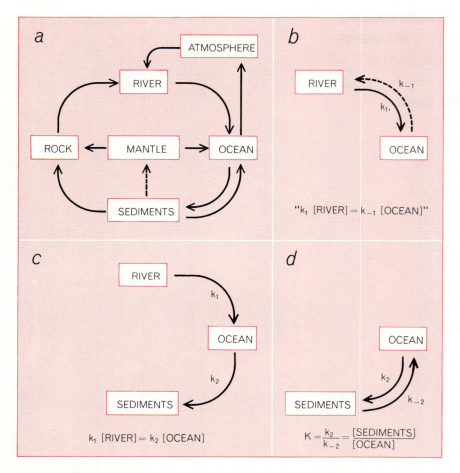

GRAND GEOCHEMICAL CYCLE (*a*) summarizes the global pathways taken sooner or later by the three-score elements that pass through the ocean and maintain its saltiness. The three "thalassochemical" models (*b, c, d*) abstracted from it are more helpful when trying to understand the rate laws governing the transport of specific elements. The rate constants, *k*, are expressed as a fraction: one over some number of years. The brackets enclose concentrations of the element being studied, specified according to its environment. The "cyclic" model (*b*) accounts for 90 percent of the chloride in river water. Its rate law is in quotation marks because extra factors, such as the area of the ocean, must be incorporated. The "steady state" model (*c*) works well for reactive trace metals; the reciprocal of k_2 is simply the residence time in the ocean. The "equilibrium" model (*d*) seems the most appropriate for the hydrogen ion (H$^+$) and the ions of the major metals, such as sodium.

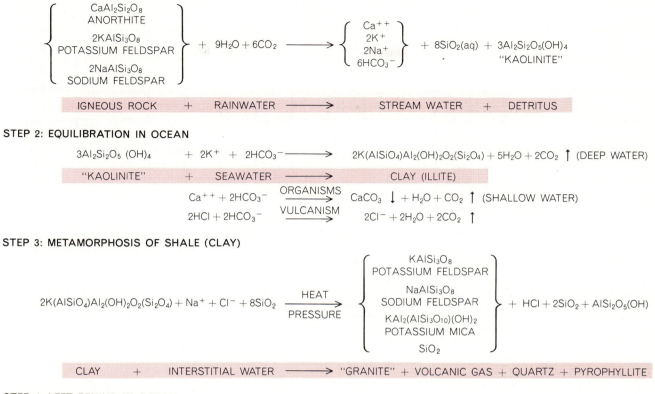

STEP 1: WEATHERING OF IGNEOUS ROCK

$$\left\{\begin{array}{c} \text{CaAl}_2\text{Si}_2\text{O}_8 \\ \text{ANORTHITE} \\ 2\text{KAlSi}_3\text{O}_8 \\ \text{POTASSIUM FELDSPAR} \\ 2\text{NaAlSi}_3\text{O}_8 \\ \text{SODIUM FELDSPAR} \end{array}\right\} + 9\text{H}_2\text{O} + 6\text{CO}_2 \longrightarrow \left\{\begin{array}{c} \text{Ca}^{++} \\ 2\text{K}^+ \\ 2\text{Na}^+ \\ 6\text{HCO}_3{}^- \end{array}\right\} + 8\text{SiO}_2(\text{aq}) + 3\text{Al}_2\text{Si}_2\text{O}_5(\text{OH})_4$$
"KAOLINITE"

IGNEOUS ROCK + RAINWATER ⟶ STREAM WATER + DETRITUS

STEP 2: EQUILIBRATION IN OCEAN

$$3\text{Al}_2\text{Si}_2\text{O}_5(\text{OH})_4 + 2\text{K}^+ + 2\text{HCO}_3{}^- \longrightarrow 2\text{K}(\text{AlSiO}_4)\text{Al}_2(\text{OH})_2\text{O}_2(\text{Si}_2\text{O}_4) + 5\text{H}_2\text{O} + 2\text{CO}_2 \uparrow \text{ (DEEP WATER)}$$

"KAOLINITE" + SEAWATER ⟶ CLAY (ILLITE)

$$\text{Ca}^{++} + 2\text{HCO}_3{}^- \xrightarrow{\text{ORGANISMS}} \text{CaCO}_3 \downarrow + \text{H}_2\text{O} + \text{CO}_2 \uparrow \text{ (SHALLOW WATER)}$$

$$2\text{HCl} + 2\text{HCO}_3{}^- \xrightarrow{\text{VULCANISM}} 2\text{Cl}^- + 2\text{H}_2\text{O} + 2\text{CO}_2 \uparrow$$

STEP 3: METAMORPHOSIS OF SHALE (CLAY)

$$2\text{K}(\text{AlSiO}_4)\text{Al}_2(\text{OH})_2\text{O}_2(\text{Si}_2\text{O}_4) + \text{Na}^+ + \text{Cl}^- + 8\text{SiO}_2 \xrightarrow[\text{PRESSURE}]{\text{HEAT}} \left\{\begin{array}{c} \text{KAlSi}_3\text{O}_8 \\ \text{POTASSIUM FELDSPAR} \\ \text{NaAlSi}_3\text{O}_8 \\ \text{SODIUM FELDSPAR} \\ \text{KAl}_2(\text{AlSi}_3\text{O}_{10})(\text{OH})_2 \\ \text{POTASSIUM MICA} \\ \text{SiO}_2 \end{array}\right\} + \text{HCl} + 2\text{SiO}_2 + \text{AlSi}_2\text{O}_5(\text{OH})$$

CLAY + INTERSTITIAL WATER ⟶ "GRANITE" + VOLCANIC GAS + QUARTZ + PYROPHYLLITE

STEP 4: LEFT BEHIND IN OCEAN

$$\text{Na}^+ + \text{Cl}^-$$

ONLY SALT REMAINS after the ocean "laboratory" has finished processing the complex of chemicals removed from igneous rock by rainwater containing dissolved carbon dioxide. Step 1 yields a solution of alkali ions and bicarbonate ($\text{HCO}_3{}^-$) ions in which hydrated silica (SiO_2) and aluminosilicate detritus are suspended. In crystalline form the aluminosilicate would be kaolinite. In the ocean (*Step 2*) the "kaolinite" is complexed with potassium ions (K^+) to form illite clay. Marine organisms use the calcium ion (Ca^{++}) to make calcium carbonate shells, which form sediments in shallow water. Hydrochloric acid (HCl), injected by undersea volcanoes, reacts with bicarbonate ions, returning some carbon dioxide to the atmosphere. In Step 3 clay is metamorphosed into "granite." Sodium chloride (*Step 4*) remains. Although some of this sequence is hypothetical, something very similar seems to take place.

avid for *d* electrons would be determined by the oxidation potential of seawater, which is a measure of its ability to extract electrons from a substance just as its *p*H is a measure of its ability to extract protons. The oxidation potential of seawater has the high value of .75 volt, enabling it to extract the maximum possible number of electrons from nearly all elements except the noble metals (platinum group) and the halogens (fluorine family).

Surprisingly, however, the oxidation potential of seawater does not seem to control the oxidation states of many metals that have partially filled *d* shells. One reason is that most reactions proceed by a mechanism in which only a single electron is transferred at a time. Such transfers occur most readily when the reactants are adsorbed on surfaces where atomic geometry and electric-charge distribution are able to expedite the redistribution of electrons (hence the utility of catalysts, which provide such surfaces). But surfaces of any kind are few and far between in the ocean, and (with the exception of manganese nodules) those that do exist are poor catalysts. A second reason for the failure of the sea's oxidation potential to control valence states is that organisms sometimes excrete electron-rich substances, which then remain in that reduced state in spite of seawater's apparent capacity to oxidize them.

Manganese nodules are porous chunks of metallic oxides up to several centimeters in diameter, widely distributed over the ocean floor. They evidently exist because they are autocatalytic for the reaction that produces them. Because of their porous structure, nodules have a surface area of as much as 100 square meters per gram. The autocatalytic property seems to extend to an entire suite of metals that coprecipitate with manganese: iron, cobalt, nickel, copper, zinc, chromium, vanadium, tungsten and lead. Nodules found on the flanks of oceanic ridges contain significant concentrations of metals, such as nickel, that are scarce in seawater itself. This suggests that the nodules are collecting juvenile metals as the metals leak from the mantle at the fissure of the ridge. One would like to know why the nodule metals are present in oxide form rather than, as one would expect, in carbonate form.

The level of the discussion so far might best be called thalassopoetry. The discussion can be made more serious in two ways. One approach—the "geochemical balance"—has employed a computer to follow in detail as many as 60 elements as they move through the geochemical cycle, from igneous rock back

← NO d-ORBITALS → ← ———— PARTIALLY FILLED d-ORBITALS ———— → ← FULL d-ORBITALS →

OH^-																	He
Li^+	Be	$B(OH)_3$										HCO_3^-	NO_3^-	O_2	F^-		Ne
Na^+	Mg^{2+}	$Al(OH)_3$										$Si(OH)_4$	HPO_4^{2-}	SO_4^{2-}	Cl^-		Ar
K^+	Ca^{2+}	Sc	$Ti(OH)_4$	VO_2^+	CrO_4^{2-}	$Mn(OH)_3$	$Fe(OH)_3$	$CoCl^+$	Ni^{2+}	$CuCl^+$	Zn^{2+}	Ga	$Ge(OH)_4$	$HAsO_4^{2-}$	SeO_4^{2-}	Br^-	Kr
Rb^+	Sr^{2+}	Y	Zr	Nb	MoO_4^{2-}	Tc	Ru	Rh	Pd	$AgCl_3^{2-}$	$CdCl_2$	In	Sn	$Sb(OH)_6^-$	Te	I^- / IO_3^-	Xe
Cs^+	Ba^{2+}	RARE EARTHS 3+	Hf	Ta	WO_4^{2-}	Re	Os	Ir	Pt	$AuCl_2^-$	$HgCl_3^-$	Tl^+	$Pb(OH)^+$	BiO^+	Po	At	Rn
Fr^+	Ra^{2+}	Ac	Th	Pa	$UO_2 (CO_3)_3^{4-}$												

☐ MAJOR ELEMENTS ☐ MINOR ELEMENTS ☐ TRACE ELEMENTS ☐ DETECTED ☐ UNDETECTED

PERIODIC TABLE, as prepared by the "thalassochemist," shows the form in which the detectable elements appear in seawater. In each box the element normally found in that place in the usual periodic table is shown in color; the elements associated with it are in black. Thus carbon appears predominantly as HCO_3^-, arsenic as $HAsO_4^{2-}$ and so on. The superscripts show the number of positive or negative charges carried by each ion. Iodine's two forms, I^- and IO_3^-, are about equally common. Except for the noble gases (*last column at right*), all the elements dissolved in the sea must be present as ions. When an element (other than a noble gas) is shown by itself, without a plus or minus charge, it means that its preferred ionic form in seawater is not yet established.

to metamorphosed sediments. In the second approach the actual chemistry of each element is followed by applying the thermodynamic methods of Josiah Willard Gibbs to systems regarded as being near equilibrium. This effort was launched by Lars Gunnar Sillén of Sweden and has been pursued by Robert M. Garrels of Northwestern University and by Heinrich D. Holland of Princeton University.

Of course no chemist in his right mind would talk seriously about equilibria in a system of variable temperature, pressure and composition that was poorly stirred, had variable inputs and contained living creatures. On the other hand, the observed uniformity of the ocean and the long periods available for reacting suggest that at least the major components are sufficiently close to equilibrium to make an investigation worthwhile. (We *know* the minor constituents are not in equilibrium.)

The equilibrium approach is based on Gibbs's phase rule, which states that the number of phases (P) possible in a system of C components at equilibrium is given by the equation $P = C + 2 - F$, where F is the number of "degrees of freedom," or quantities that may be independently varied without changing the number of phases or their composition (although F may change their relative proportions). The 2 enters the equation because only two variables, temperature and pressure, are important in most chemical reactions.

One of Sillén's most comprehensive ocean models has nine components: water, hydrochloric acid, silica, three hydroxides (aluminum, sodium and potassium), carbon dioxide and the oxides of magnesium and calcium. Observation of sea-floor sediments, aided by laboratory studies, suggests that a nine-phase ocean will result [*see illustrations on opposite page*]. If C and P both equal nine, the phase rule states that the number of degrees of freedom (F) must equal two. Logically these are temperature (which can vary over the oceanic range from −2 degrees Celsius to 30 degrees) and the chloride ion concentration (which can shift over the normal oceanic range without changing the composition of the stable phases).

A diagrammatic view of how the nine components sort themselves into phases is shown in the bottom illustration on the opposite page. Note that the liquid phase contains ions not listed either as components or as phases (for example H^+ and OH^-). Thermodynamics need not consider them explicitly because they do not vary independently; their concentrations are fixed by the equilibrium constants that connect the observed phases. Thus $H_2O = H^+ + OH^-$. Moreover, one knows that the product of H^+ and OH^- is a thermodynamic constant, which equals 10^{-14} mole per liter. Similar relations tie the entire system into a comprehensible whole, so that when all the calculations are performed one has discovered the equilibrium concentra-

tions of five cations (H^+, Na^+, K^+, Mg^{++} and Ca^{++}) and four anions (Cl^-, OH^-, HCO_3^- and CO_3^{--}).

It may seem peculiar to discuss an "atmosphere" containing only water vapor and carbon dioxide. One could easily add oxygen and nitrogen to the list of components. Since they would add no new phases, they would raise the number of degrees of freedom from two to four ($9 = 11 + 2 - 4$). The two new F's would be the total atmospheric pressure and the ratio of oxygen to nitrogen. In the study of the ocean, however, the partial pressure due to carbon dioxide is more significant than the total pressure of the atmosphere. Moreover, the presence of gaseous oxygen and nitrogen has little importance for the inorganic environment of the ocean, so that it is simpler to omit them and just as "real."

Suppose now we perturb the equilibrium of the model ocean by assuming that a submerged volcano has suddenly released enough hydrochloric acid (HCl) to double the amount of chloride ion (Cl^-). The dissociation of hydrochloric acid releases enough H^+ ions to raise the total number of hydrogen ions in the ocean from the former equilibrium value of 10^{-8} mole per liter to $10^{+.3}$. This excess of hydrogen ions almost immediately pushes all the available carbonate ions (CO_3^{--}) to bicarbonate ions (HCO_3^-) and the latter to carbonic acid (H_2CO_3). These shifts, however, only slightly depress the pH, which remains high until the slow circulation of the ocean brings the hydrogen ions in direct contact with the clay sediments on the sea floor.

The structure of clay is such that oxygen atoms at the free corners of polyhedrons carry unsatisfied negative charges, which attract positive ions [*see top illustration on next page*]. Because the ocean is so rich in sodium ions (Na^+), they occupy most of the corners of clay polyhedrons. When the excess hydrogen ions come in contact with the clay, they quickly replace the sodium ions and set them adrift. This fast reaction is limited in scope because the surface and inter-layer ion-exchange capacity of clay is not very great. Much more capacity is provided when the structure of the clay is rearranged; for example, the conversion of montmorillonite to kaolinite also consumes hydrogen atoms and releases sodium. Given sufficient time—centuries—such rearrangements inexorably take place, and the *p*H of the ocean slowly drifts back to its equilibrium value. The charge on the excess chloride introduced by the volcano will then be balanced not by H^+ but by Na^+. This slow equilibration mechanism can be regarded as the ocean's "*p*H-stat" (in analogy with "thermostat"). This clay-cation model suggests that the *p*H of the ocean has been constant over the span of geologic time and that hence the carbon dioxide content of the atmosphere has been held within narrow limits.

If only the *p*H-stat were available for leveling surges in *p*H, the ocean might be subjected to violent local fluctuations. For fast response *p*H control is taken over by a carbonate buffer system [*see bottom illustration on next page*]. In fact, until recently oceanographers neglected the clay-cation reactions and assumed that the carbonate-buffer system almost completely determined the *p*H of the ocean.

One might think that if the carbon dioxide content of the atmosphere were to decrease, carbon dioxide would flow from the sea into the atmosphere, leading to a general depletion of all carbonate species in the ocean and eventually to the dissolution of some carbonate sediments. In actuality something quite different happens because the carbonate system is its own source of hydrogen ions. Removal of carbon dioxide from water reduces the concentration of carbonic acid (H_2CO_3), the hydrated form of carbon dioxide. Replacement of this acid from bicarbonate ions requires a hydrogen ion, which can only be obtained by converting another bicar-

COMPONENTS (C)	PHASES (P)	VARIABLES (F)
H_2O	1 GAS	TEMPERATURE
HCl	2 LIQUID	Cl^-
SiO_2	3 QUARTZ (SiO_2)	
$Al(OH)_3$	4 KAOLINITE (t-o CLAY)	
NaOH	5 MONTMORILLONITE (Na-t-o-t CLAY)	
KOH	6 ILLITE (K-t-o-t CLAY)	
MgO	7 CHLORITE (Mg-t-o-t CLAY)	
CO_2	8 CALCITE ($CaCO_3$)	
CaO	9 PHILLIPSITE (Na-K FELDSPAR)	

NINE MAJOR COMPONENTS IN SEA can, to a first approximation, be combined into nine distinctive phases to satisfy the "phase rule" that governs systems in equilibrium. The rule, formulated in the 19th century by Josiah Willard Gibbs, prescribes the number of phases C, components C and degrees of freedom F in such a system: $P = C + 2 - F$. When the number of phases and components are equal, the number of degrees of freedom, F, must be two, which allows both the temperature and the chloride-ion concentration to vary without altering the number of phases. In the clay-containing phases (*4, 5, 6, 7*) the letter "t" stands for a tetrahedral crystal structure; the letter "o" stands for an octahedral structure.

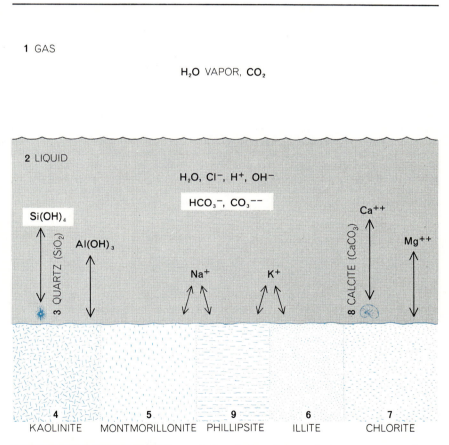

EQUILIBRIUM OCEAN MODEL, consisting of nine phases and nine components, shows how the principal constituents of the ocean distribute themselves among the atmosphere, the ocean and the sediments. Three of the constituents (HCO_3^-, CO_3^{--} and $Si(OH)_4$) are not included among nine listed components but appear as equilibrium products of those that are listed, as do seven ions (H^+, K^+, Na^+, Ca^{++}, Mg^{++}, Cl^-, OH^-). Two of the solids are shown as biological "precipitates": "quartz" (*3*) in the form of silicate structures built by radiolarians and "calcite" (*8*) in the form of calcium carbonate chambers built by foraminifera. The method of precipitation is unimportant as long as the product is stable. The equilibrium model goes far to explain why the ocean has the composition it does.

bonate ion to carbonate. The overall reaction is $2HCO_3^- \rightarrow H_2CO_3 + CO_3^{--}$. Thus instead of dissolving existing sediments, removing carbon dioxide from the sea may actually precipitate carbonate. This reaction can be seen in the "whitings" of the sea over the Bahama Banks,

where cold deep water, rich in dissolved carbon dioxide and calcium, is forced to the surface and warmed. As carbon dioxide escapes into the air, the pH drops and aragonite ($CaCO_3$) precipitates, turning large areas of the ocean white with a myriad of small crystals.

The reaction above conserves charge, which means that the "alkalinity"—the traditional name for the concentration of sodium ion ("alkali") needed to balance this negative charge—is also conserved. The "carbonate alkalinity," defined as the bicarbonate concentration plus twice the carbonate concentration, is useful because it remains fixed even when the relative amounts of the two species vary.

The system can be visualized with the help of the illustration on the opposite page, which is the "Bjerrum plot" for carbonic acid at constant alkalinity. It takes its name from Niels Bjerrum, who introduced such plots in 1914; it shows the interrelations between the various compounds in the world carbonate system as a function of pH. Although the diagram ignores variations of pressure, temperature and salinity, it displays the essential features of the system.

The Bjerrum plot facilitates a semiquantitative discussion of the relation of atmospheric carbon dioxide to oceanic carbon dioxide. Over the next 20 years we shall burn enough fossil fuel to double the amount of carbon dioxide in the atmosphere from 320 parts per million to 640. On the plot this is indicated by shifting the line A, corresponding to 320 parts per million, to position B, 640 parts per million.

To produce this shift some 2.5×10^{18} grams of carbon dioxide must be added to the atmosphere. If the altered atmosphere were to come to equilibrium with the ocean, the pH of the ocean would drop from its present value of 8.15 to 7.89—still well within the range tolerated by marine organisms. This cannot happen, however, because the total mass of carbon dioxide in the ocean (Σ in the Bjerrum plot) plus the carbon dioxide in the atmosphere would have to increase from its present value, 128.9×10^{18} grams, to 138.3×10^{18} grams. The difference, 9.4×10^{18} grams, is nearly four times the amount added to the atmosphere.

The long-term equilibration process for such an atmospheric doubling can be broken down into two steps. First the pH-buffer system operates: 2.5×10^{18} grams, or 2 percent of the total mass, is added to the world system at constant alkalinity. The result of this step is the line C in the diagram, corresponding to a total mass of 131.4×10^{18} grams, an atmospheric carbon dioxide content of 390 parts per million and an oceanic pH of 8.08. Next, if the ocean has time to equilibrate with its sediments, the pH-stat will operate, returning the system to pH 8.15 at a constant total mass. The re-

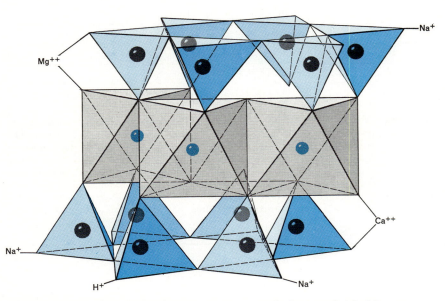

THREE-LAYER CLAY PARTICLE has a layer of octahedrons sandwiched between two layers of tetrahedrons. Each octahedron consists of an atom of aluminum surrounded by six closely packed atoms of oxygen. Each tetrahedron consists of a silicon atom surrounded by four atoms of oxygen. The polyhedrons are tied into layers at shared corners where a single oxygen atom is bonded to a silicon atom on one side and to an aluminum atom on the other. At the free corners the oxygen atoms bear unsatisfied negative charges that attract cations such as sodium (Na+) and potassium (K+). If the hydrogen-ion concentration should rise in the vicinity of clay, free hydrogen ions tend to be exchanged for sodium ions, which are released. In addition, many doubly charged metal ions can replace Si^{4+} at the centers of tetrahedrons and Si^{4+} can replace Al^{3+} in the octahedrons. Whenever this occurs, another cation is bound to the structure to conserve charge. Such reactions apparently exert considerable control over the ocean's composition and hydrogen-ion concentration.

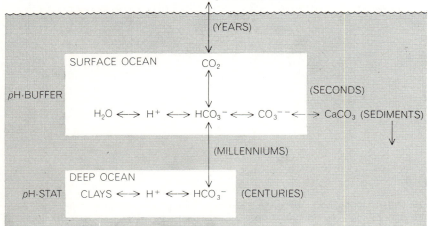

HYDROGEN-ION CONCENTRATION, or pH, of the ocean is controlled by two mechanisms, one that responds swiftly and one that takes centuries. The first, the "pH-buffer," operates near the surface and maintains equilibrium among carbon dioxide, bicarbonate ion (HCO_3^-), carbonate ion (CO_3^{--}) and sediments. The slower mechanism, the "pH-stat," seems to exert ultimate control over pH; it involves the interaction of bicarbonate ions and protons (H^+) with clays. Clay will accept protons in exchange for sodium ions (primarily).

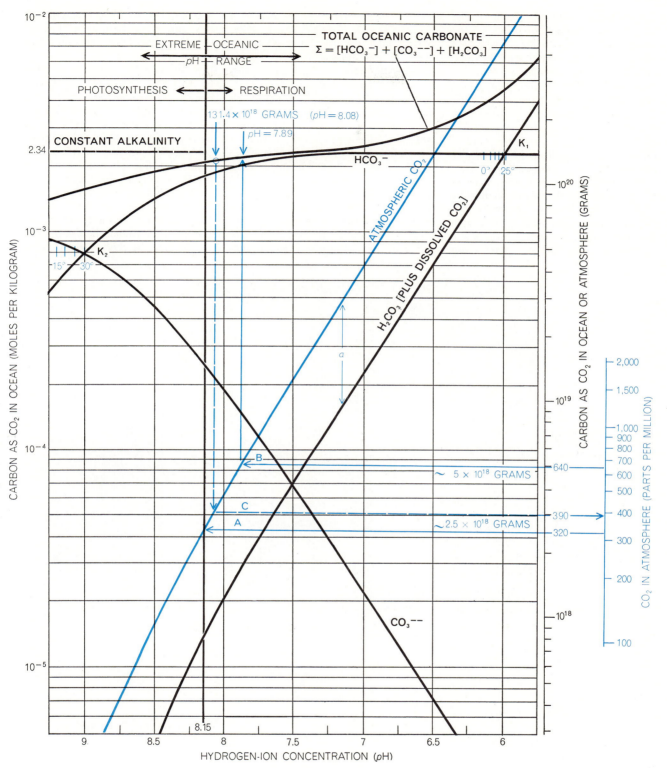

OCEANIC CARBONATE SYSTEM can be represented by a "Bjerrum diagram" that shows how carbonate in its several forms varies with the ocean's pH, or hydrogen-ion concentration. The diagram is plotted for a constant "carbonate alkalinity" of 2.34×10^{-3} moles of carbonate per kilogram of seawater (scale at left). "System point" K_1 shows where the concentrations of bicarbonate ion (HCO_3^-) and carbonic acid (H_2CO_3) are equal. At K_2 the concentrations of bicarbonate and carbonate (CO_3^{--}) are equal. The exact locations of K_1 and K_2 are shown for a range of temperatures (in degrees Celsius) at constant conditions of salinity and pressure. The top curve, Σ, is the sum of oceanic carbonate in all its forms. The normal pH of the ocean is 8.15. The two short arrows at top mark the normal biological limits: at 7.95 the available oxygen has been consumed by respiration; at 8.35 photosynthesis has removed so much carbon dioxide that absorption from the atmosphere rises sharply. The limits of oceanic pH lie between 7.45 and 8.6. The amount of carbon dioxide in the atmosphere (colored curve and scale at far right) is related to the amount of carbon dioxide dissolved in the ocean by alpha (α), the average worldwide solubility of carbon dioxide in seawater. The consequences of doubling the carbon dioxide in the atmosphere from 320 parts per million (A) to 640 parts (B) are discussed in the text of the article, as is line C.

sult of this step is that the alkalinity rises by 2 percent, which in terms of the Bjerrum plot means that the system will return to normal except that all the numbers on the concentration axes will be multiplied by 1.02. The long-range effect of a sudden doubling of the atmosphere's carbon dioxide, therefore, is to increase the ultimate value 2 percent, from 320 parts per million to 326, and some of that increase will ultimately find its way into vegetation and humus.

It is obvious that rates are crucial in the global distribution of carbon dioxide. The wind-stirred surface layer of the sea exchanges carbon dioxide rapidly with the atmosphere, requiring less than a decade for equilibration. Because this layer is only about 100 meters deep it contains only a tiny fraction of the ocean's total volume. Large-scale disposal of atmospheric carbon dioxide therefore requires that the gas be dissolved and transported to deep water.

Such vertical transport takes place almost exclusively in the Weddell Sea off the coast of Antarctica. Every winter, when the Weddell ice shelf freezes, the salt excluded from the newly formed ice increases the salinity and hence the density of the water below. This ice-cold water, capable of containing more dissolved gas than an equal volume of tropical water, cascades gently down the slope of Antarctica to begin a 5,000-year journey northward across the bottom of the ocean. The carbon dioxide in this "antarctic bottom water" has plenty of time to come to equilibrium with clay sediments.

Enough fossil fuel has been burned in the past century to have raised the carbon dioxide content of the atmosphere from about 290 parts per million to 350 parts. Since the actual level is now 320 parts per million, about half of the carbon dioxide put into the air has been removed. Although proof is lacking, a principal removal agent is undoubtedly antarctic bottom water. The process is so slow, however, that the carbon dioxide content of the atmosphere may reach 480 parts per million before the end of the century. By then it should be clear

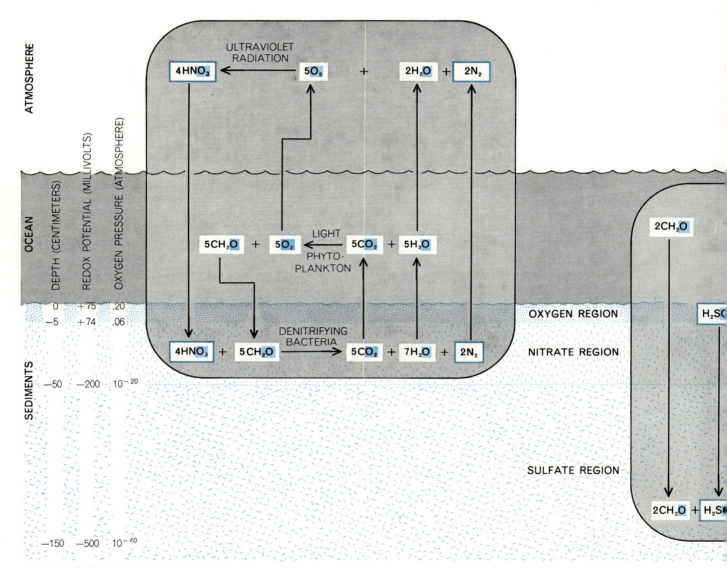

BACTERIA IN MARINE SEDIMENTS, although scarce by terrestrial soil standards, play a major role in replenishing the oxygen of the atmosphere and in limiting the accumulation of organic sediments. The bacteria concerned are buried in fine-grained sediments from several centimeters to several tens of centimeters below the ocean floor, with limited access to free oxygen for respiration. Thus deprived, they use the oxygen in nitrates and sulfates to oxidize organic compounds, represented by CH_2O. The actual reactions are far more complex than indicated here. The net result, however, is that denitrifying bacteria (*left*) release free nitrogen and convert carbon to a form (carbon dioxide) that can be reutilized by phytoplankton. These organisms, in turn, release free oxy-

if man's inadvertent global experiment (altering the atmosphere's carbon dioxide content) will have the predicted effect of changing the earth's climate. In principle an increase in atmospheric carbon dioxide should reduce the amount of long-wavelength radiation sent back into space by the earth and thus produce a greenhouse effect, slightly raising the average world temperature.

Having described an equilibrium model of the ocean that neglected the atmosphere's content of nitrogen and oxygen, I should not leave the reader with the impression that the continued presence of these two gases in the atmosphere is independent of the ocean. If

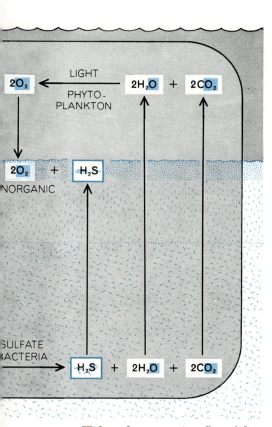

gen. Without the cooperative effort of these two groups of organisms the oxygen in the atmosphere might all be fixed by high-energy processes within some 10 million years. The sulfate bacteria (right) play a role in the recycling of sulfur and oxygen.

the ocean were truly in equilibrium with the atmosphere, it would long since have captured all the atmospheric oxygen in the form of nitrates, both in solution and in sediments. This catastrophe has apparently been averted by the intervention of certain marine bacteria that have the happy faculty of releasing nitrogen gas from nitrate compounds and of converting the oxygen to a form that can later be liberated by phytoplankton.

The story is this. A variety of high-energy processes in the atmosphere continuously break the triple chemical bond that holds two nitrogen atoms together in a nitrogen molecule (N_2). The bonds can be broken by ultraviolet photons, by cosmic rays, by lightning and by the explosions in internal-combustion engines. Once dissociated, nitrogen atoms can react with oxygen to form various oxides, which are then carried to the ground by rainfall. In the soil these oxides are useful as fertilizer. Ultimately large amounts of them reach the sea. They do not, however, accumulate there and no one is really sure why.

The best guess is that denitrifying bacteria in oceanic sediments use the oxygen of nitrate to oxidize organic molecules when they run out of free oxygen [see left half of illustration on these two pages]. The nitrogen is released directly as a gas, which goes into solution but is available for return to the atmosphere. The oxygen emerges in molecules of water and carbon dioxide. The carbon dioxide is assimilated by phytoplankton, which build the carbon into organic compounds and release the oxygen as dissolved gas, also available for return to the atmosphere. Without these coupled biological processes the atmospheric fixation of nitrogen would probably exhaust the world's oxygen supply in less than 10 million years. Nevertheless, the amount of nitrogen returned to the atmosphere from the sediments is so small that we may never be able to measure it directly: the yearly return is less than one two-thousandth of the total nitrogen dissolved in the sea.

Another little-known epicycle in the global oxygen cycle probably has the effect of limiting the net accumulation of carbon in the form of oil-bearing shale, tar sands and petroleum. After denitrifying bacteria have consumed the nitrate in young sediments, sulfate bacteria begin oxidizing organic matter with the oxygen contained in sulfates [see right half of illustration on these two pages]. The product, in addition to water and carbon dioxide, is hydrogen sulfide, the foul-smelling compound that character-

izes environments deficient in oxygen. In undisturbed mud the hydrogen sulfide never reaches the surface because it is inorganically reoxidized to sulfate as soon as it comes in contact with free oxygen. It seems likely that the bacterial turnover of oxygen in sulfate is so rapid that half of the world's oxygen passes through this epicycle in about 50,000 years.

The global activities of man have now reached such a scale that they are beginning to have a profound effect on marine chemistry and biology. We are learning that even the ocean is not large enough to absorb all the waste products of industrial society. The experiment involving the release of carbon dioxide is now in progress. DDT, only 25 years on the scene, is now found in the tissues of animals from pole to pole and has pushed several species of birds close to extinction. The concentration of lead in plants, animals and man has increased tenfold since tetraethyl lead was first used as an antiknock agent in motor fuels. And high levels of mercury in fish have forced the abandonment of some commercial fisheries. (Lead and mercury are systemic enzyme poisons.) Of the total petroleum production some .2 percent gets slopped into the sea in half a dozen major accidents each year. (At least six of the rare gray whales died last year after migrating through the oil slick off Santa Barbara caused by the blowout of a well casing belonging to the Union Oil Company.) Conceivably a persistent oil film could change the surface reflectivity of the ocean enough to alter the world's energy balance. The rapid increase in the use of nitrogen fertilizers leaves a nitrate excess that runs into rivers, lakes and ultimately reaches the sea. The sea can probably tolerate the runoff indefinitely but along the way the nitrogen creates algal "blooms" that are hastening the dystrophication of lakes and estuaries.

It is fashionable today to view the ocean as the last global frontier, waiting only technological "development." Thermodynamically it is easier to extract fresh water from sewage than from seawater. Ecologically it is wiser to keep our concentrated nutrients on land than to dilute them beyond recall in the ocean. Sociologically, and probably economically, it makes more sense to process our junkyards for usable metals than to mine the deep-sea floor. The task is to persuade our engineers and business companies that working with sewage and junk is just as challenging as oceanography and thalassochemistry.

7

Ocean Waves

by Willard Bascom
August 1959

*Men have always been fascinated, and sometimes
awed, by the rhythmic motions of the sea's surface.
A century of observation and experiment has
revealed much about how these waves are generated
and propagated.*

Man is by nature a wave-watcher. On a ship he finds himself staring vacantly at the constant swell that flexes its muscles just under the sea's surface; on an island he will spend hours leaning against a palm tree absently watching the rhythmic breakers on the beach. He would like to learn the ways of the waves merely by watching them, but he cannot, because they set him dreaming. Try to count a hundred waves sometime and see.

Waves are not always so hypnotic. Sometimes they fill us with terror, for they can be among the most destructive forces in nature, rising up and overwhelming a ship at sea or destroying a town on the shore. Usually we think of waves as being caused by the wind, because these waves are by far the most common. But the most destructive waves are generated by earthquakes and undersea landslides. Other ocean waves, such as those caused by the gravitational attraction of the sun and the moon and by changes in barometric pressure, are much more subtle, often being imperceptible to the eye. Even such passive elements as the contour of the sea bottom, the slope of the beach and the curve of the shoreline play their parts in wave action. A wave becomes a breaker, for example, because as it advances into increasingly shallow water it rises higher and higher until the wave front grows too steep and topples forward into foam and turbulence. Although the causes of this beautiful spectacle are fairly well understood, we cannot say the same of many other aspects of wave activity. The questions asked by the wave-watcher are nonetheless being answered by intensive studies of the sea and by the examination of waves in large experimental tanks. The new knowledge has made it possible to measure the power and to forecast the actions of waves for the welfare of those who live and work on the sea and along its shores.

Toss a pebble into a pond and watch the even train of waves go out. Waves at sea do not look at all like this. They are confused and irregular, with rough diamond-shaped hillocks and crooked valleys. They are so hopelessly complex that 2,000 years of observation by seafarers produced no explanation beyond the obvious one that waves are somehow raised by the wind. The description of the sea surface remained in the province of the poet who found it "troubled, unsettled, restless. Purring with ripples under the caress of a breeze, flying into scattered billows before the torment of a storm and flung as raging surf against the land; heaving with tides breathed by a sleeping giant."

The motions of the oceans were too complex for intuitive understanding. The components had to be sorted out and dealt with one at a time. So the first theoreticians cautiously permitted a perfect train of waves, each exactly alike, to travel endlessly across an infinite ocean. This was an abstraction, but it could at least be dealt with mathematically.

Early observers noticed that passing waves move floating objects back and forth and up and down, but do not transport them horizontally for any great distance. From the motion of seaweeds the motion of the water particles could be deduced. But it was not until 1802 that Franz Gerstner of Germany constructed the first wave theory. He showed that water particles in a wave move in circular orbits. That is, water at the crest moves horizontally in the direction the wave is going, while in the trough it moves in the opposite direction. Thus each water particle at the surface traces a circular orbit, the diameter of which is equal to the height of the wave [see *illustration page 64*]. As each wave passes, the water returns almost to its original position. Gerstner observed that the surface trace of a wave is approximately a trochoid: the curve described by a point on a circle as it rolls along the underside of a line. His work was amplified by Sir George Airy later in the 19th century, by Horace Lamb of England in the present century, and by others.

The first wave experimentalists were Ernst and Wilhelm Weber of Germany, who in 1825 published a book on studies employing a wave tank they had invented. Their tank was five feet long, a foot deep and an inch wide, and it had glass sides. To make waves in the tank they sucked up some of the fluid through a tube at one end of it and allowed the fluid to drop back. Since the Weber brothers experimented not only with water and mercury but also with brandy, their persistence in the face of temptation has been an inspiration to all subsequent investigators. They discovered that waves are reflected without loss of energy, and they determined the shape of the wave surface by quickly plunging in and withdrawing a chalk-dusted slate. By watching particles suspended in the water they confirmed the theory that water particles move in a circular orbit, the size of which diminishes with depth. At the bottom, they observed, these orbits tend to be flattened.

As increasingly bolder workers contributed ideas in the 20th century, many of the complexities of natural waves found their way into equations. However, these gave only a crude, empirical answer to the question of how wind energy is transferred to waves. The necessity for the prediction of waves and surf for amphibious operations in World War II attracted the attention of Harald U.

DIFFRACTION OF OCEAN WAVES is clearly visible in this aerial photograph of Morro Bay, Calif. The waves are diffracted as they pass the end of the lower jetty. Variations in the way the waves break are caused by contours of the shore and the bottom.

Sverdrup and Walter Munk of the Scripps Institution of Oceanography. As a result of their wartime studies of the interaction of winds and waves they were the first investigators to give a reasonably complete quantitative description of how wind gets energy into the waves. With this description wave studies seemed to come of age, and a new era of research was launched.

Let us follow waves as they are generated at sea by the wind, travel for perhaps thousands of miles across the ocean and finally break against the shore. The effectiveness of the wind in making waves is due to three factors: its average velocity, the length of time it blows and the extent of the open water across which it blows (called the fetch).

Waves and the Wind

Waves start up when the frictional drag of a breeze on a calm sea creates ripples. As the wind continues to blow, the steep side of each ripple presents a surface against which the moving air can press directly. Because winds are by nature turbulent and gusty, wavelets of all sizes are at first created. The small, steep ones break, forming whitecaps, releasing some of their energy in turbulence and possibly contributing part of it to larger

waves that overtake them. Thus as energy is added by the wind the smaller waves continually give way to larger ones which can store the energy better. But more small waves are continually formed, and in the zone where the wind moves faster than the waves there is a wide spectrum of wavelengths. This is the generating area, and in a large storm it may cover thousands of square miles. If storm winds apply more force than a wave can accept, the crest is merely steepened and blown off, forming a breaking wave at sea. This happens when the wave crest becomes a wedge of less than 120 degrees and the height of the wave is about a seventh of its length. Thus a long wave can accept more energy from the wind and rise much higher than a short wave passing under the same wind. When the wind produces waves of many lengths, the shortest ones reach maximum height quickly and then are destroyed, while the longer ones continue to grow.

A simple, regular wave-train can be described by its period (the time it takes two successive crests to pass a point), by its wavelength (the distance between crests) and by its height (the vertical distance between a trough and a succeeding crest). Usually, however, there are several trains of waves with different

wavelengths and directions present at the same time, and their intersection creates a random or a short-crested diamond pattern. Under these conditions no meaningful dimensions can be assigned to wave period and length. Height, however, is important, at least to ships; several crests may coincide and add their heights to produce a very large wave. Fortunately crests are much more likely to coincide with troughs and be canceled out. There is no reason to believe that the seventh wave, or some other arbitrarily numbered wave, will be higher than the rest; that is a myth of the sea.

Since waves in a sea are so infinitely variable, statistical methods must be employed to analyze and describe them. A simple way to describe height, for example, is to speak of significant height—the average height of the highest third of the waves. Another method, devised in 1952 by Willard J. Pierson, Jr., of New York University, employs equations like those that describe random noise in information theory to predict the behavior of ocean waves. Pierson superposes the regular wave-trains of classical theory in such a way as to obtain a mathematically irregular pattern. The result is most conveniently described in terms of energy spectra. This scheme assigns a value for the square of the wave height to each

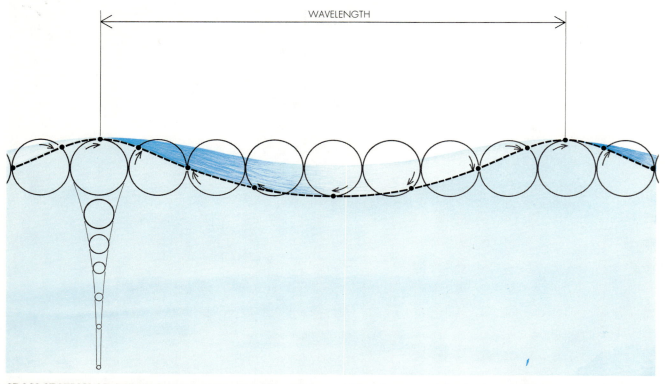

CROSS SECTION OF OCEAN WAVE traveling from left to right shows wavelength as distance between successive crests. The time it takes two crests to pass a point is the wave period. Circles are orbits of water particles in the wave. At the surface their diameter equals the wave height. At a depth of half the wavelength (*left*), orbital diameter is only 4 per cent of that at surface.

frequency and direction. Then, by determining the portion of the spectrum in which most of the energy is concentrated, the average periods and lengths can be obtained for use in wave forecasting.

Over a long fetch, and under a strong, steady wind, the longer waves predominate. It is in such areas of sea that the largest wind waves have been recorded. The height of the waves in a train does not, however, bear any simple relationship to their other two dimensions: the period and the wavelength. The mariner's rule of thumb relates wave height to wind velocity and says that the height ordinarily will not be greater than half the wind speed. This means that an 80-mile-per-hour hurricane would produce waves about 40 feet high.

The question of just how large individual waves at sea can actually be is still unsettled, because observations are difficult to make and substantiate from shipboard in the midst of a violent storm. Vaughan Cornish of England spent half a century collecting data on waves, and concluded that storm waves over 45 feet high are rather common. Much higher waves have been fairly well authenticated on at least two occasions.

In October, 1921, Captain Wilson of the 12,000-ton S.S. *Ascanius* reported an extended storm in which the recording barometer went off the low end of the scale. When the ship was in a trough on an even keel, his observation post on the ship was 60 feet above the water level, and he was certain that some of the waves that obscured the horizon were at least 10 feet higher than he was, accounting for a total height of 70 feet or more. Commodore Hayes of the S.S. *Majestic* reported in February, 1923, that his ship had experienced winds of hurricane force and waves of 80 feet in height. Cornish examined the ship, closely interrogated the officers and concluded that waves 60 to 90 feet high, with an average height of 75 feet, had indeed been witnessed.

A wave reported by Lieutenant Commander R. P. Whitemarsh in the *Proceedings of the U. S. Naval Institute* tops all others. On February 7, 1933, the U.S.S. *Ramapo*, a Navy tanker 478 feet long, was en route from Manila to San Diego when it encountered "a disturbance that was not localized like a typhoon . . . but permitted an unobstructed fetch of thousands of miles." The barometer fell to 29.29 inches and the wind gradually rose from 30 to 60 knots over several days. "We were running directly downwind and with the sea. It would have been disastrous to have steamed on any other course." From among a number of separately determined observations, that of the watch officer on the bridge was selected as the most accurate. He declared that he "saw seas astern at a level above the mainmast crow's-nest and at the moment of observation the horizon was hidden from view by the waves approaching the stern." On working out the geometry of the situation from the ship's plan, Whitemarsh found that this wave must have been at least 112 feet high [*see illustration at the bottom of the next two pages*]. The period of these waves was clocked at 14.8 seconds and their velocity at 55 knots.

As waves move out from under the winds that raise them, their character changes. The crests become lower and more rounded, the form more symmetrical, and they move in trains of similar period and height. They are now called swell, or sometimes ground swell, and in this form they can travel for thousands of miles to distant shores. Happily for mathematicians, swell coincides much more closely with classical theory than do the waves in a rough sea, and this renews their faith in the basic equations.

Curiously enough, although each wave moves forward with a velocity

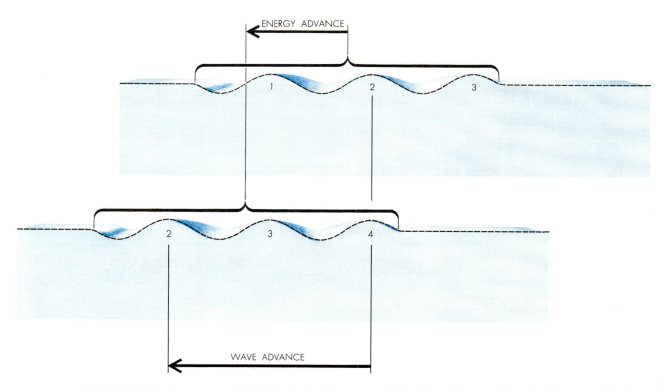

MOVING TRAIN OF WAVES advances at only half the speed of its individual waves. At top is a wave train in its first position. At bottom the train, and its energy, have moved only half as far as wave 2 has. Meanwhile wave 1 has died, but wave 4 has formed at the rear of the train to replace it. Waves arriving at shore are thus remote descendants of waves originally generated.

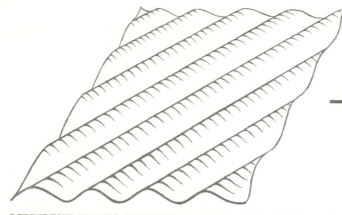

DIFFERENT TRAINS OF WAVES, caused by winds of different directions and strengths, make up the surface of a "sea." The vari-ous trains, three of which are represented diagrammatically here, have a wide spectrum of wavelengths, heights and directions. When

that corresponds to its length, the energy of the group moves with a velocity only half that of the individual waves. This is because the waves at the front of a group lose energy to those behind, and gradually disappear while new waves form at the rear of the group. Thus the composition of the group continually changes, and the swells at a distance are but remote descendants of the waves created in the storm [*see illustration on preceding page*]. One can measure the period at the shore and obtain from this a correct value for the wave velocity; however, the energy of the wave train traveled from the storm at only half that speed.

Waves in a swell in the open ocean are called surface waves, which are defined as those moving in water deeper than half the wavelength. Here the bottom has little or no effect on the waves because the water-particle orbits diminish so rapidly with depth that at a depth of half the wavelength the orbits are only 4 per cent as large as those at the surface. Surface waves move at a speed in miles per hour roughly equal to 3.5 times the period in seconds. Thus a wave with a period of 10 seconds will travel about 35 miles per hour. This is the average period of the swell reaching U. S. shores, the period being somewhat longer in the Pacific than the Atlantic. The simple relationship between period and wavelength (length $=5.12T^2$) makes it easy to calculate that a 10-second wave will have a deep-water wavelength of about 512 feet. The longest period of swell ever reported is 22.5 seconds, which corresponds to a wavelength of around 2,600 feet and a speed of 78 miles per hour.

Waves and the Shore

As the waves approach shore they reach water shallower than half their

wavelength. Here their velocity is controlled by the depth of the water, and they are now called shallow-water waves. Wavelength decreases, height increases and speed is reduced; only the period is unchanged. The shallow bottom greatly modifies the waves. First, it refracts them, that is, it bends the wave fronts to approximate the shape of the underwater contours. Second, when the water becomes critically shallow, the waves break [*see illustration on page 71*].

Even the most casual observer soon notices the process of refraction. He sees that the larger waves always come in nearly parallel to the shoreline, even though a little way out at sea they seem to be approaching at an angle. This is the result of wave refraction, and it has considerable geological importance because its effect is to distribute wave energy in such a way as to straighten coastlines. Near a headland the part of the wave front that reaches shallow water first is slowed down, and the parts of

it in relatively deep water continue to move rapidly. The wave thus bends to converge on the headland from all sides. As it does, the energy is concentrated in less length of crest; consequently the height of the crest is increased. This accounts for the old sailors' saying: "The points draw the waves."

Another segment of the same swell will enter an embayment and the wave front will become elongated so that the height of the waves at any point along the shore is correspondingly low. This is why bays make quiet anchorages and exposed promontories are subject to wave battering and erosion—all by the same waves. One can deal quantitatively with this characteristic of waves and can plot the advance of any wave across waters of known depths. Engineers planning shoreline structures such as jetties or piers customarily draw refraction diagrams to determine in advance the effect of waves of various periods and direction. These diagrams show successive

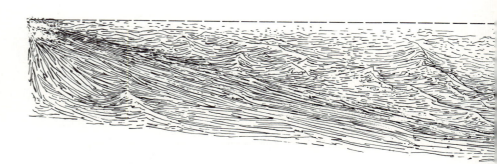

WAVE 112 FEET HIGH, possibly the largest ever measured in the open sea, was encountered in the Pacific in 1933 by the U.S.S. *Ramapo*, a Navy tanker. This diagram shows

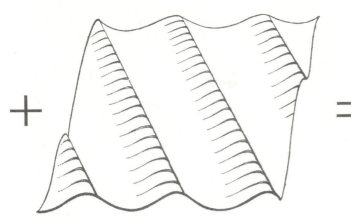 + =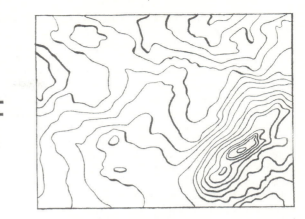

they meet, the result is apparent confusion, represented at far right by a topographic diagram drawn from actual photographs of

the sea surface. The pattern becomes so complex that statistical methods must be used to analyze the waves and predict their height.

positions of the wave front, partitioned by orthogonals into zones representing equal wave energy [*see illustration on next page*]. The ratio of the distances between such zones out at sea and at the shore is the refraction coefficient, a convenient means of comparing energy relationships.

Refraction studies must take into account surprisingly small underwater irregularities. For example, after the Long Beach, Calif., breakwater had withstood wave attack for years, a short segment of it was suddenly wrecked by waves from a moderate storm in 1930. The breakwater was repaired, but in 1939 waves breached it again. A refraction study by Paul Horrer of the Scripps Institution of Oceanography revealed that long-period swell from exactly 165 degrees (southsoutheast), which was present on only these two occasions, had been focused at the breach by a small hump on the bottom, 250 feet deep and more than seven miles out at sea. The hump had acted as

a lens to increase the wave heights to 3.5 times average at the point of damage.

During World War II it was necessary to determine the depth of water off enemy-held beaches against which amphibious landings were planned. Our scientists reversed the normal procedure for refraction studies; by analyzing a carefully timed series of aerial photographs for the changes in length (or velocity) and direction of waves approaching a beach, they were able to map the underwater topography.

The final transformation of normal swell by shoal or shallow water into a breaker is an exciting step. The waves have been shortened and steepened in the final approach because the bottom has squeezed the circular orbital motion of the particles into a tilted ellipse; the particle velocity in the crest increases and the waves peak up as they rush landward. Finally the front of the crest is unsupported and it collapses into the trough. The wave has broken and the

orbits exist no more. The result is surf.

If the water continues to get shallower, the broken wave becomes a foam line, a turbulent mass of aerated water. However, if the broken wave passes into deeper water, as it does after breaking on a bar, it can form again with a lesser height that represents the loss of energy in breaking. Then it too will break as it moves into a depth critical to its new height.

The depth of water beneath a breaker, measured down from the still-water level, is at the moment of breaking about 1.3 times the height of the breaker. To estimate the height of a breaker even though it is well offshore, one walks from the top of the beach down until the crest of the breaking wave is seen aligned with the horizon. The vertical distance between the eye and the lowest point to which the water retreats on the face of the beach is then equal to the height of the wave.

The steepness of the bottom influences

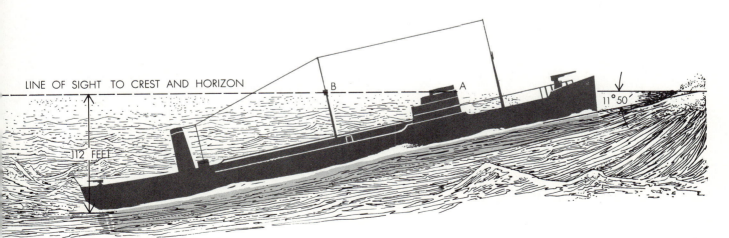

how the great wave was measured. An observer at A on the bridge was looking toward the stern and saw the crow's-nest at B in his

line of sight to crest of wave, which had just come in line with horizon. From geometry of situation, wave height was calculated.

the character of the breakers. When a large swell is forced by an abrupt underwater slope to give up its energy rapidly, it forms plunging breakers—violent waves that curl far over, flinging the crest into the trough ahead. Sometimes, the air trapped by the collapsing wave is compressed and explodes with a great roar in a geyser of water [see illustration on opposite page]. However, if the bottom slope is long and gentle, as at Waikiki in Hawaii, the crest forms a spilling breaker, a line of foam that tumbles down the front of the partly broken wave as it continues to move shoreward.

Since waves are a very effective mechanism for transporting energy against a coast, they are also effective in doing great damage. Captain D. D. Gaillard of

the U. S. Army Corps of Engineers devoted his career to studying the forces of waves on engineering structures and in 1904 reported some remarkable examples of their destructive power. At Cherbourg, France, a breakwater was composed of large rocks and capped with a wall 20 feet high. Storm waves hurled 7,000-pound stones over the wall and moved 65-ton concrete blocks 60 feet. At Tillamook Rock Light off the Oregon coast, where severe storms are commonplace, a heavy steel grating now protects the lighthouse beacon, which is 139 feet above low water. This is necessary because rocks hurled up by the waves have broken the beacon several times. On one occasion a rock weighing 135 pounds was thrown well above the

lighthouse-keeper's house, the floor of which is 91 feet above the water, and fell back through the roof to wreck the interior.

At Wick, Scotland, the end of the breakwater was capped by an 800-ton block of concrete that was secured to the foundation by iron rods 3.5 inches in diameter. In a great storm in 1872 the designer of the breakwater watched in amazement from a nearby cliff as both cap and foundation, weighing a total of 1,350 tons, were removed as a unit and deposited in the water that the wall was supposed to protect. He rebuilt the structure and added a larger cap weighing 2,600 tons, which was treated similarly by a storm a few years later. There is no record of whether he kept his job

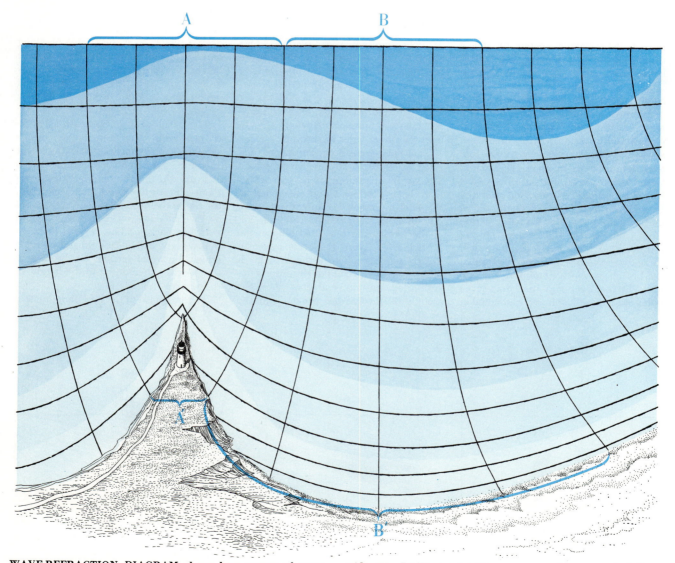

WAVE-REFRACTION DIAGRAM shows how energy of wave front at A is all concentrated by refraction at A′ around small headland area. Same energy at B enters a bay but is spread at beach over wide area B′. Horizontal lines are wave fronts; vertical lines divide energy into equal units for purposes of investigation. Such studies are vital preliminaries to design of shoreline structures.

and tried again. Gaillard's computations show that the wave forces must have been 6,340 pounds per square foot.

Tsunamis

Even more destructive than wind-generated waves are those generated by a sudden impulse such as an underwater earthquake, landslide or volcano. A man-made variation of the sudden impulse is the explosion of nuclear bombs at the surface of the sea, which in recent years have become large enough to be reckoned with as possible causes of destructive waves.

The public knows such waves as tidal waves, although they are in no way related to the tides and the implication has long irritated oceanographers. It was proposed that the difficulty could be resolved by adopting the Japanese word *tsunami*. Some time later it was discovered that Japanese oceanographers are equally irritated by this word; in literal translation tsunami means tidal wave! However, tsunami has become the favored usage for seismic sea waves.

Like the plunger in a wave channel, the rapid motion or subsidence of a part of the sea bottom can set a train of waves in motion. Once started, these waves travel great distances at high velocity with little loss of energy. Although their height in deep water is only a few feet, on entering shallow water they are able to rise to great heights to smash and inundate shore areas. Their height depends almost entirely on the configuration of the coastline and the nearby underwater contours. Tsunamis have periods of more than 15 minutes and wavelengths of several hundred miles. Since the depth of water is very much less than half the wavelength, they are regarded as long- or shallow-water waves, even in the 13,000-foot average depth of the open ocean, and their velocity is limited by the depth to something like 450 miles per hour.

These fast waves of great destructive potential give no warning except that the disturbance that causes them can be detected by a seismograph. The U. S. Coast Guard operates a tsunami warning network in the Pacific that tracks all earthquakes, and when triangulation indicates that a quake has occurred at sea, it issues alerts. The network also has devices to detect changes in wave period which may indicate that seismic waves are passing [see the article "Tsunamis," by Joseph Bernstein; SCIENTIFIC AMERICAN, August 1954]. Curiously the influence of the system may not be entirely

WAVE-CREATED "GEYSER" results when large breakers smash into a very steep beach. They curl over and collapse, trapping and compressing air. This compressed air then explodes as shown here, with spray from a 12-foot breaker leaping 50 feet into the air.

beneficial. Once when an alert was broadcast at Honolulu, thousands of people there dashed down to the beach to see what luckily turned out to be a very small wave.

Certain coasts near zones of unrest in the earth's crust are particularly prone to such destructive waves, especially the shores of the Mediterranean, the Caribbean and the west coast of Asia. On the world-wide scale, they occur more frequently than is generally supposed: nearly once a year.

A well-known seismic sea wave, thoroughly documented by the Royal Society of London, originated with the eruption of the volcano Krakatoa in the East Indies on August 27, 1883. It is not certain whether the waves were caused by the submarine explosion, the violent movements of the sea bottom, the rush of water into the great cavity, or the

dropping back into the water of nearly a cubic mile of rock, but the waves were monumental. Their period close to the disturbance was two hours, and at great distances about one hour. Waves at least 100 feet high swept away the town of Merak, 33 miles from the volcano; on the opposite shore the waves carried the man-of-war *Berow* 1.8 miles inland and left it 30 feet above the level of the sea. Some 36,380 people died by the waves in a few hours. Tide gauges in South Africa (4,690 miles from Krakatoa), Cape Horn (7,820 miles) and Panama (11,470 miles) clearly traced the progress of a train of about a dozen waves, and showed that their speed across the Indian Ocean had been between 350 and 450 miles per hour.

A tsunami on April 1, 1946, originating with a landslide in the Aleutian submarine trench, produced similar effects,

HUNDRED-FOOT "TIDAL WAVE," or tsunami, wrought impressive destruction at Scotch Cap, Alaska, in 1946. Reinforced concrete lighthouse that appears in top photograph was demolished, as shown in lower photograph, which was made from a higher angle. Atop the plateau a radio mast, its foundation 103 feet above sea, was also knocked down. Lighthouse debris was on plateau. Same tsunami, started by an Aleutian Island earthquake, hit Hawaiian Islands, South America and islands 4,000 miles away in Oceania.

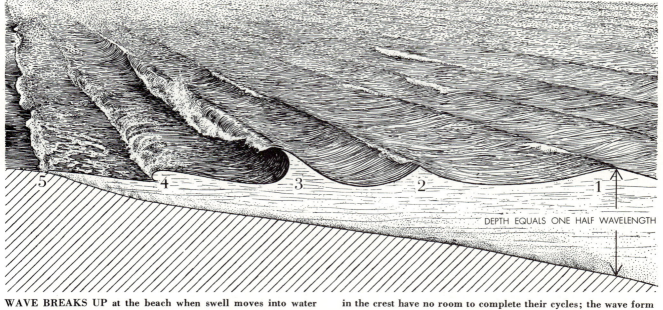

WAVE BREAKS UP at the beach when swell moves into water shallower than half the wavelength (1). The shallow bottom raises wave height and decreases length (2). At a water depth 1.3 times the wave height, water supply is reduced and the particles of water in the crest have no room to complete their cycles; the wave form breaks (3). A foam line forms and water particles, instead of just the wave form, move forward (4). The low remaining wave runs up the face of the beach as a gentle wash called the uprush (5).

fortunately on less-populated shores. It struck hard at the Hawaiian Islands, killing several hundred people and damaging property worth millions of dollars. At Hilo, Hawaii, the tsunami demonstrated that such waves are virtually invisible at sea. The captain of a ship standing off the port was astonished upon looking shoreward to see the harbor and much of the city being demolished by waves he had not noticed passing under his ship. The same waves caused considerable damage throughout the islands of Oceania, 4,000 miles from epicenter, and on the South American coast, but they were most spectacular at Scotch Cap in Alaska. There a two-story reinforced-concrete lighthouse marked a channel through the Aleutian Islands. The building, the base of which was 32 feet above sea level, and a radio mast 100 feet above the sea were reduced to bare foundations by a wave estimated to be more than 100 feet high [see illustration on opposite page].

Uncontrollable geologic disturbances will cause many more seismic sea waves in the future, and since the world's coastal population is continuously increasing, the greatest wave disaster is yet to come. Within the next century we can expect that somewhere a wave will at least equal the one that swept the shores of the Bay of Bengal in 1876, leaving 200,000 dead.

Tides and Other Waves

The rhythmic rise and fall of the sea level on a coast indicate the passage of a true wave we call a tide. This wave is driven, as almost everyone knows, by the gravitational influence of the sun and the moon. As these bodies change their relative positions the ocean waters are attracted into a bulge that tends to remain facing the moon as the earth turns under it; a similar bulge travels around the earth on the opposite side. The wave period therefore usually corresponds to half the lunar day.

When the sun and the moon are aligned with the earth, the tides are large (spring tides); when the two bodies are at right angles with respect to the earth, the tides are small (neap tides). By using astronomical data it is possible to predict the tides with considerable accuracy. However, the height and time of the tide at any place not on the open coast are primarily a function of the shape and size of the connection to the ocean.

Still another form of wave is a seiche, a special case of wave reflection. All enclosed bodies of water rock with characteristics related to the size of the basin. The motion is comparable to the sloshing of water in the bathtub when one gets out quickly. In an attempt to return to stability the water sways back and forth with the natural period of the tub (mine has a period of two seconds). Similarly a tsunami or a barometric pressure-change will often set the water in a bay rocking as it passes. In fact, the tsunami itself may reflect back and forth across the ocean as a sort of super-seiche.

In addition to seiches, tides, tsunamis and wind waves there are other waves in the sea. Some travel hundreds of feet beneath the surface along the thermocline, the interface between the cold deep water and the relatively warm surface layer. Of course these waves cannot be seen, but thermometers show that they are there, moving slowly along the boundary between the warm layer and the denser cold water. Their study awaits proper instrumentation. Certain very low waves, with periods of several minutes, issue from storms at sea. These long-period "forerunners" may be caused by the barometric pulsation of the entire storm against the ocean surface. Since they travel at hundreds of miles an hour, they could presumably be used as storm warnings or storm-center locators. Other waves, much longer than tides, with periods of days or weeks and heights of less than an inch, have been discovered by statistical methods and are now an object of study.

The great advances both in wave theory and in the actual measurement of waves at sea have not reduced the need for extensive laboratory studies. The solution of the many complex engineering problems that involve ships, harbors, beaches and shoreline structures requires that waves be simulated under ideal test conditions. Such model studies in advance of expensive construction permit much greater confidence in the designs.

Experimental Tanks

The traditional wave channel in which an endless train of identical small waves is created by an oscillating plunger is still in use, but some of the new wave tanks are much more sophisticated. In some the channel is covered, so that a high velocity draft of air may simulate the wind in making waves. In others, like the large tank at the Stevens Institute of Technology in Hoboken, New Jersey, artificial irregular waves approach the variability of those in the deep ocean. In such tanks proposed ship designs, like those of the America's Cup yacht *Columbia,* are tested at model size to see how they will behave at sea.

The ripple tank, now standard apparatus for teaching physics, has its place in shoreline engineering studies for conveniently modeling diffraction and refraction. Even the fast tsunamis and the very slow waves of the ocean can be modeled in the laboratory. The trick is to use layers of two liquids that do not mix, and create waves on the interface between them. The speeds of the waves can be controlled by adjusting the densities of the liquids.

To reduce the uncertainties in extrapolation from the model to prototype, some of the new wave tanks are very large. The tank of the Beach Erosion Board in Washington, D.C. (630 feet long and 20 feet deep, with a 500-horsepower generator), can subject quarter-scale models of ocean breakwaters to six-foot breakers. The new maneuvering tank now under construction at the David Taylor Model Basin in Carderock, Md., measures 360 by 240 feet, is 35 feet deep along one side and will have wave generators on two sides that can independently produce trains of variable waves. Thus man can almost bring the ocean indoors for study.

The future of wave research seems to lie in refinement of the tools for measuring, statistically examining and reproducing in laboratories the familiar wind waves and swell as well as the more recently discovered varieties. It lies in completing the solution of the problem of wave generation. It lies in the search for forms of ocean waves not yet discovered—some of which may exist only on rare occasions. Nothing less than the complete understanding of all forms of ocean waves must remain the objective of these studies.

Fog

by Joel N. Myers
December 1968

A kind of grounded cloud, fog can halt sea, air and highway travel. When combined with air pollutants, it can be lethal. Ways are known, however, not only to dissipate fog but also to inhibit its formation.

Fog, once little more than a nuisance except at sea, has become an important hazard to modern man. Its effects on travel are intensified by the speed of the airplane and the automobile. Dense fogs close airports in the U.S. for an average of 115 hours per year, and in 1967 they cost the nation's airlines an estimated $75 million in disrupted schedules as well as inestimable inconvenience to passengers. On present-day turnpikes fog can be disastrous; a single pileup on a fog-shrouded freeway in Los Angeles involved more than 100 vehicles. Above all, fog in combination with air pollution now increasingly afflicts large cities. Its potential was suggested alarmingly by the London smog of December, 1952.

On December 5 a dense fog settled over the city. A strong inversion—warm air lying above cold air—blocked the removal of polluted air by vertical movement, and at the same time the winds were so light that such air was only gradually removed by horizontal movement. Within 24 hours the tons of smoke, dust and chemical fumes given off by the city's furnaces, factories and automobiles turned the fog brown and then black. Two days later the visibility in the city had been reduced to a matter of inches. People fell off wharfs into the Thames and drowned. Others wandered blindly until they died of exposure to the cold. The toxic air afflicted millions of Londoners with smarting eyes, coughing, nausea and diarrhea. It was estimated that the smog killed at least 4,000 persons, mainly from respiratory disorders, and caused permanent injury to tens of thousands.

The principal source of toxicity in most pathological smogs appears to be sulfur dioxide from the smokestacks. The sulfur dioxide is oxidized in the air to sulfur trioxide, which in turn combines with fog droplets to form sulfuric acid. One therefore inhales sulfuric acid with each breath, with resulting acute irritation of the throat and lungs. This seems to have been the chief cause of injury and death in the London smog and in the 1948 smog in Donora, Pa., which killed 20 victims and sickened nearly half of the 14,000 inhabitants.

Strictly speaking, the perennial "smogs" of Los Angeles are misnamed, as they frequently consist of a haze of pollutants that are trapped by an inversion layer of dry air rather than by fog. Polluted air, however, can generate fog and cause it to persist. Once formed, the fog reflects solar radiation back into space, thereby providing an environment favorable to the accumulation of further pollution. In effect fog, itself partly a product of pollution, causes pollution to increase.

Some metropolitan areas in the U.S. suffer from light smogs up to 100 days of the year, and during fall and winter dangerous smog conditions can become frequent. The long-term effects of the urban smogs are believed to include chronic bronchitis, emphysema, asthma and lung cancer. The polluted air corrodes metals, rots wood, causes paint to discolor and flake and may cause extensive damage to vegetation and livestock. In addition to the directly damaging effects, there are indications that the combination of fog and air pollution may be upsetting the delicate balance of man's ecosystem. Smog plays an important role in influencing the earth's gain and loss of radiation, which ultimately determines most of the climatic variables. Smog reduces sunlight, lowers daytime temperature and wind speed, raises humidity and is even suspected of causing a decrease in rainfall. Thus for many reasons—the hazards to travel, the damaging effects of smog, the modification of climate—fog is a growing problem that will demand increasing research attention in the years ahead.

Fog is simply a cloud on the ground, composed, like any cloud, of tiny droplets of water or, in rare cases, of ice crystals, forming an ice fog. Ice fogs usually occur only in extremely cold climates, because the water droplets in a cloud are so tiny they do not solidify until the air temperature is far below freezing, generally 30 degrees below zero Celsius or lower. The droplets of fogs are nearly spherical; they vary in diameter between two and 50 microns and in concentration between 20 and 500 droplets per cubic centimeter of air. The transparency of a fog depends mainly on the concentration of droplets; the more droplets, the denser the fog. A wet sea fog may contain a gram of water per cubic meter; a very light fog may have as little as .02 gram of water per cubic meter.

Since water is 800 times denser than air, investigators were long puzzled as to why fogs did not quickly disappear through fallout of the water particles to the ground. Even allowing for air resistance, a 10-micron water droplet falls in still air with a velocity of .3 centimeter per second, and a 20-micron droplet falls at 1.3 centimeters per second. To explain the persistence of fogs many early investigators concluded that the droplets must be hollow (that is, bubbles). It turns out, however, that the droplets are fully liquid and do fall at the predictable rate, but in fog-creating conditions they either are buoyed up by rising air currents or are continually replaced by new droplets condensing from the water vapor in the air.

The atmosphere always contains

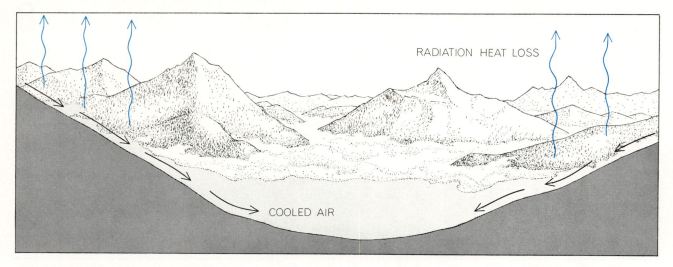

FOG IS FORMED when moist air is cooled; the air then cannot hold as much moisture in the form of water vapor as it can when it is warmer. As the humidity approaches the saturation point tiny water droplets form, obscuring visibility. The diagram indicates how a radiation fog, one of three kinds of "cooling fog," forms. Night cooling has reduced the air's temperature and cause water droplets to appear. The cold, fog-filled air drains downhill and accumulates in low-lying areas (*bottom illustration on page 78*).

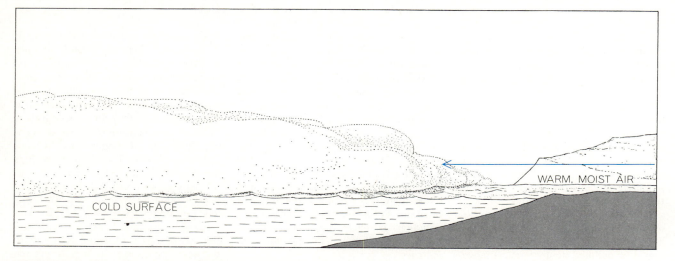

ADVECTION FOG, the second kind of cooling fog, forms when warm, moist air is carried across a cold surface and is cooled to the saturation point. Advection fogs are common at sea where warm air masses come in contact with cold ocean currents (*see top illustration on pages 76-77*) and on land during the cold months when moist tropical air masses are carried across chilled grounds.

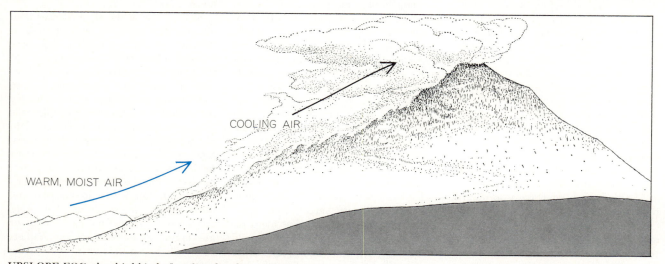

UPSLOPE FOG, the third kind of cooling fog, forms when an air mass is forced upward. In the diagram a topographic barrier forces the air mass to rise. As pressure diminishes the air mass expands, grows cooler and, as in both other cases, soon becomes saturated.

some water vapor, supplied by evaporation from bodies of water, vegetation and other sources. Since the air's capacity for holding water in the form of vapor decreases with falling temperature, even comparatively dry air will reach the saturation point—100 percent relative humidity—when it is cooled sufficiently. At that point the vapor of course begins to condense into liquid water. Fogs often form, however, at humidities well below 100 percent. The droplets condense on tiny particles of dust in the air called condensation nuclei. These are hygroscopic particles, which, because of their affinity for water vapor, initiate condensation at subsaturation humidities—sometimes as low as 65 percent. The nucleus on which the water condenses, which may be a soil particle or a grain of sea salt, a combustion product or cosmic dust, usually dissolves in the droplet. Because the saturation point is lower for solutions than it is for pure water, the droplets of solution tend to condense more water vapor on them and grow in size. A rise in the air's humidity will also enlarge the droplets and will form more of them, thereby thickening a light fog into a dense one.

Given suitable conditions of temperature and humidity, the density of a fog and its microphysical properties will depend on the availability of condensation nuclei and their nature. Fogs become particularly dense near certain industrial plants because of the high concentration of hygroscopic combustion particles in the air. This is not to say that air pollution is generally a primary cause of fog formation, but it does cause fogs to form sooner and persist longer, and it makes them dirtier—hence less transparent—than fogs that develop at higher humidities in relatively clean air.

From a meteorological point of view fogs are classified in several types according to the gross natural processes that generate them. Over land the most common type is the "radiation fog" that arises from nighttime cooling of the earth's surface and the lower atmosphere. As the earth radiates away its heat during the night, fog may form if the air in contact with the cooling ground is moist or, even though it is fairly dry at first, if it is cooled a great deal. Radiation fogs occur most frequently over swampy terrain and in deep, narrow valleys where cold air draining down from the hillsides concentrates in the valley bottom. The likelihood of fog formation depends considerably on the wind speed. A moderate to strong wind, by moving the air about and diluting the cooling effect,

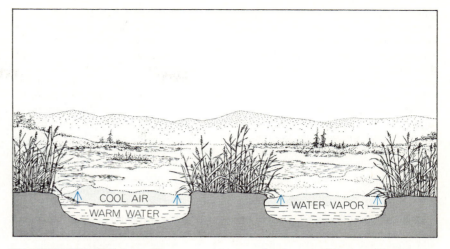

WARM-WATER FOG forms like steam over water that is covered by a much colder air mass. Water vapor from the comparatively warm water rises into the colder air and is rapidly condensed into fog droplets. The fog's intensity depends on the temperature differential.

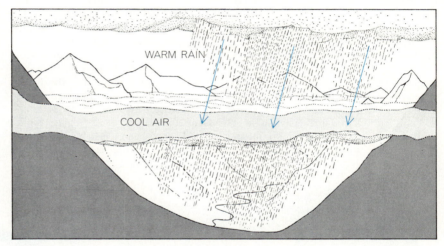

WARM-RAIN FOG is formed when raindrops from higher clouds encounter a layer of cooler air near the ground; evaporating raindrops saturate the cold air layer. As the diagram indicates, the fog looks like a low cloud to an observer looking up from the valley floor.

tends to prevent the formation of fog. If the air is calm, only a thin layer of air next to the ground is much affected by the cooling, so that the condensation may be restricted to dew or a shallow ground fog. The condition most likely to produce an extensive fog is a slight breeze; by generating turbulence near the ground such a breeze may spread out the cooled surface air to form a layer of fog several hundred feet deep. Above this cold layer the air remains warm.

One might suppose that radiation fogs should reach a maximum around dawn, when the air temperature ordinarily is at its daily low point. Actually it has been found that this type of fog sometimes becomes thickest shortly after sunrise. The reason appears to be that the sun's early rays, not yet strong enough to evaporate the fog droplets, generate turbulence that intensifies and thickens the

fog layer. The fog does not begin to dissipate until the sun is high enough to heat the atmosphere, stirring the foggy air so that it mixes with the warm, dry air above. Obviously in pockets that are topographically shielded from the sun the fog will remain longer.

A somewhat different process produces what are called advection fogs. In this case the fog arises from the movement of humid air over a surface that is already cold. Most sea fogs are of this type; indeed, the foggiest places in the world are the areas above cold ocean currents. Advection also commonly plays a part, in combination with nocturnal cooling, in the generation of land fogs. In a different way air moving up a mountainside sometimes produces fog: the air expands and cools as it rises because the atmospheric pressure diminishes with altitude. Advection and upslope fogs can provide additions to the water supply. On some coastal mountains in California, for example, the ground receives more water from fog dripping off vegetation than it does from rain. Residents of some arid regions take advantage of this phenomenon by suspending arrays of nylon threads to extract water from drifting fogs.

Fog is also produced by the familiar steaming process we observe above a hot bath or a heated kettle or on a hot roof or parking lot after a summer shower; the vapor rising from the warm water quickly condenses into steam in the cooler air above. In this way the evaporation from a body of water on an unseasonably cold night may generate a shallow fog (up to perhaps 50 feet). Warm rain falling through cool air also can give rise to a steamlike fog; it is a common cause of the fogging in of airfields during rainy weather.

The principal natural agents of fog dispersal are sunshine and brisk wind, but a dense fog has built-in resistance to the sun. Because fog is an excellent reflector of sunlight, only 20 to 40 percent of the impinging solar energy penetrates it to warm the ground and the foggy air, and only part of that heat is available to evaporate the fog droplets. Consequently a dense fog strongly resists dissipation

by the sun. Moreover, the resistance is increased when the fog becomes a thick smog.

To what extent has our era of industrialism intensified the fog problem? This is very difficult to determine, because the meteorological records are incomplete and the phenomenon itself is highly variable. The incidence and intensity of fogs vary widely with time and season in any given place and from one place to another, even within a few miles. Furthermore, weather-observation practices have changed considerably over the past century: the observations are much more frequent than they used to be, cover many more locations and are complicated by changes both in the location of observatories and in the observers' nomenclature. The definition of "dense" fog, for example, has been revised repeatedly by the U.S. Weather Bureau. Nevertheless, although reliable comparisons cannot be made in detail, a survey of past records indicates some general trends. It appears that the incidence, duration and probably the density of fogs vary directly with the amount of industrial activity and air pollution in the area involved.

Perhaps the most reliable set of data for a single location is the one for the

CLASSIC EXAMPLE of an advection fog is the light, low-lying bank seen obscuring

city of Prague in Czechoslovakia, where consistent observations of fogs have been made for the past century and a half. The records show that in the period since 1881 Prague has had nearly twice as many fogs as in the preceding 80-year period. In general it appears

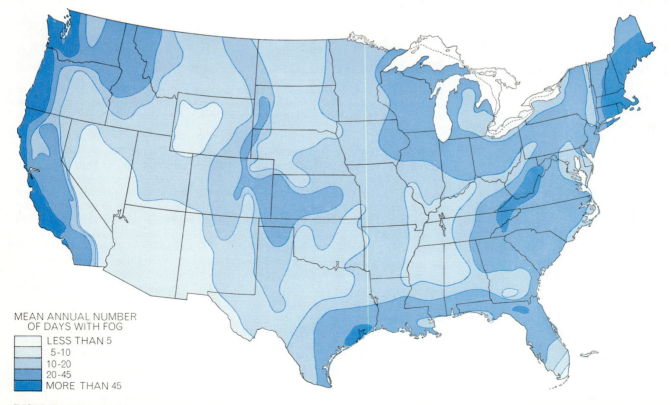

MEAN ANNUAL NUMBER
OF DAYS WITH FOG

LESS THAN 5
5-10
10-20
20-45
MORE THAN 45

DISTRIBUTION OF FOGS in the continental U.S. is shown by area shading that indicates the days of dense fog per year reported by 251 weather stations from 1900 to 1960. The least foggy parts of the mainland are the desert areas of Arizona, California and Nevada; foggiest are the Pacific and New England coasts. Appalachian hills often have rain fogs; the valleys, radiation fogs.

the Golden Gate Bridge at the entrance to San Francisco Bay in the aerial photograph. When a warm, moist air mass passes over the cold waters of the California Current, it is cooled to the saturation point and some of its water vapor is condensed into fog droplets.

that fog tends to be more frequent in and near cities than in unpolluted rural areas. The available data do not indicate, however, if fogs tend to be denser in cities than they are elsewhere. It is conceivable that the relatively warm microclimate of large cities (due to the solar heating of their streets and buildings and the urban artificial heat) may prevent the strong nocturnal cooling that is usually a prerequisite for dense fogs. If that is so, perhaps the places where dense fogs occur most frequently are highly industrialized small cities and suburban communities that lie downwind of metropolises.

A fog recently studied by Charles L. Hosler, Jr., a meteorologist at Pennsylvania State University, may serve as an illustration of the problem. The fog is seen shortly after sunrise in Bald Eagle Valley in central Pennsylvania on days when little or no fog is observed elsewhere in the state. The valley contains large industrial plants that daily discharge billions of hygroscopic particles and tons of water vapor into the atmosphere. Although fog formation is aided by the overnight drainage of cold air into Bald Eagle Valley, the pollution is a major contributing factor.

What can be done about the growing fog problem? The phenomenon has become a matter of active concern for many interests—government, industry, the airlines, the military—and it is being attacked by investigators in a wide range of disciplines. Clearly one of the prime needs is a vigorous assault on the complex problem of air pollution. Nor is pollution confined exclusively to the atmosphere. A form of pollution that is responsible for some fogs and that has received little attention is the "thermal pollution" of natural water resulting from industrial practices. It gives rise to the formation of fogs by increasing the normal rate of water evaporation.

Enormous quantities of heated water are discharged into our streams and lakes by industrial operations that use water for various cooling purposes. The principal offenders are steel mills, paper factories, sewage-treatment plants, certain chemical and manufacturing plants and particularly electric-power generating stations. Power plants based on nuclear fuel, to which many utilities are now turning, use enormous quantities of water for cooling. The power industry estimates that by 1980 it will be using one-fifth of the total free water runoff in the U.S. for cooling. Thus, ironically, nuclear plants, which are counted on to reduce air pollution by replacing plants that use fossil fuels, may become a major factor in spawning fog and thereby trapping air pollution in the form of smog.

Some industrial plants, notably in the power industry, have taken to getting rid of their heated water by evaporating it in massive cooling towers instead of dumping it into bodies of water. Unfortunately this method is expensive and may still produce fogs. Some towers evaporate hundreds of cubic meters of water per hour, and the tremendous flux of vapor, if trapped under an inversion on a windless night, could generate and maintain a dense fog over a large city. Federal aviation officials believe a cooling tower recently built three miles north of the Morgantown Municipal Airport in West Virginia is responsible for local fogs that have begun to trouble the area. And it appears that fog plumes from other cooling towers may have contributed to automobile accidents on highways near them.

Cooling towers might be turned into assets rather than liabilities in farming areas. The fog they produce could prevent heat from escaping from the soil and thus protect cold-sensitive crops against frost. Some farmers in localities of frequent fogs report that fog blankets extend the frost-free growing season and thereby increase their crop yields. Russian meteorologists have successfully used artificial fogs to protect vineyards from frost. Perhaps power companies should be encouraged to build their new plants in rural regions where farming could benefit from such frost protection.

Among the various possibilities for dealing with the fog problem the approach that has been explored most actively up to now is the idea of dissipating fog by some artificial means. Over the past several decades hundreds of schemes for doing this have been proposed and many have been tried, but so far no universally practicable method has been found. What are the prospects that an effective and not too expensive technique could be developed?

The most direct method would be

ICE FOG (*above*) fills the Nenana River valley near Fairbanks, Alaska. Ice fogs consist mainly of ice crystals that form in air cooled to 30 degrees below zero Celsius or lower. They may appear because a temperature drop has produced nearly 100 percent humidity or because the air is saturated by the addition of water vapor.

RADIATION FOG (*below*) spills out of Bald Eagle Valley through a mountain gap near Lock Haven, Pa. Radiation fogs are caused by the night cooling of the earth's surface, which reduces the air temperature and raises the humidity to near-saturation. In Bald Eagle Valley industrial wastes contribute to the fog-forming processes.

simply to blow the fog away, using an artificial wind. This tactic is actually employed around some settlements in the Arctic to disperse ice fogs. In the frigid atmosphere the water vapor emitted from human settlements (indeed, even the exhalations of a reindeer herd) can easily saturate the air and produce a local fog; at a temperature of 30 degrees or more below zero C. the fog consists mainly of ice crystals. Giant fans have been used successfully to free settlements of such fogs. Obviously, however, the method is applicable only on a small scale and in special situations.

Another attack on fog, based on the principle of evaporating the droplets by one means or another, has been applied in several interesting ways. The best-known of these is the FIDO method (Fog Investigation and Dispersal Operation), in which fuel-oil fires have been used to burn off (that is, evaporate) fog on airfields. The British resorted to this technique on military fields in World War II and successfully cleared them for more than 2,000 takeoffs and landings that could not have been undertaken otherwise. The method has important drawbacks, however. It is expensive, requiring hundreds of dollars' worth of fuel to clear a jet runway for 10 to 15 minutes; it creates a fire hazard for planes landing on the field, and it cannot dissipate all dense fogs. Moreover, the smoke and moisture released by the combustion of oil hinder evaporation of the fog. It has been proposed that this problem might be obviated by using electricity, jet-engine exhausts or anthracite coal to provide the heat, but the high cost would still remain a major objection. Drying the air with chemicals rather than heat has been effective in clearing fog in some cases; here again, however, the procedure is too expensive for wide use, and the chemicals employed tend to be corrosive.

Among the various principles of fog dispersal that have been tested, the most promising seems to be the injection of a catalyst or some other agent that will cause the droplets to coalesce and thus grow large enough to fall quickly to the ground. This type of attack has proved its worth in fogs consisting of supercooled droplets. By seeding the fog with particles of a very cold substance such as dry ice or liquid propane, one can cause some of the fog droplets to freeze. Water vapor in the air then condenses onto these ice crystals; the resultant drying of the air turns additional fog droplets back into vapor. The vapor, in turn, acceler-

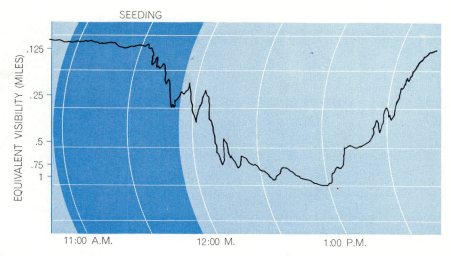

COLD-SEEDING TEST at the airport in Medford, Ore., began shortly before 11:00 A.M.; visibility, as measured by a transmissometer, was less than an eighth of a mile. Thirty minutes after the first seeding run visibility began to improve. By 12:15 P.M. it exceeded the half-mile minimum required at the airport and remained above minimum for over an hour.

ates the growth of the ice crystals, which fall to the ground as they enlarge. United Air Lines has been seeding supercooled fogs at fields in the Pacific Northwest and in Alaska for several years and has found the method to be about 80 percent successful in dissipating such fogs. The airline estimates that this investment in fog control has repaid it fivefold by maintaining the regularity of flight schedules. The method has also worked well in other areas where it is applicable. It is effective, however, only for supercooled fogs, which account for only about 5 percent of all the fogs in the U.S. Temperate Zone.

Several ideas for dispersing warm fogs by droplet coalescence are currently being explored. The Air Transport Association is sponsoring a series of tests of a chemical mixture (composition undisclosed) that is said to make droplets combine by an electrical attraction effect. It is reported to have achieved some success in dissipating radiation fogs at Sacramento, Calif., during calm or very light winds. The ability of the method to disperse moving fogs, however, remains to be demonstrated. In any attempt to control fog, the wind condition is crucial. When the air is calm, clearing it of fog presents a comparatively uncomplicated problem because the volume of air that needs to be treated is limited. On the other hand, a breeze can quickly re-fog a space that has been cleared of fog; when moderate or strong winds are blowing, it is almost impossible to maintain a clearing.

Another scheme that has been proposed for fog dispersal involves dropping carefully controlled doses of salt particles into the fog. The theory is that when these hygroscopic particles deliquesce into solution droplets, they will gain water and grow at the expense of natural fog droplets because the humidity is higher with respect to the solution than it is with respect to natural water. It is hoped that the collection of the water into fewer and larger drops will produce a rapid improvement in visibility and that the fall of the large drops to the ground will maintain the improvement for some time after seeding has ended.

For effective progress toward the economical control or modification of fogs we shall have to learn a great deal more about the basic structure of fog and the chemical, physical and electrical properties of the fog droplets. Intensive studies are going forward on various types of natural fogs and on artificial fogs produced in the laboratory. The results of these investigations will be used to test mathematical models that describe fogs in terms of such quantities as temperature, humidity, wind, condensation nuclei, concentration of droplets and amount of liquid water. It should then be possible to obtain insight into the mechanisms and energy exchanges involved in the formation and maintenance of fogs and allow meteorologists to determine which kind of fog will respond to which dispersal method.

The artificial dissipation of fog will be rather costly in any case. Much might be done to prevent the formation of fogs in the first place. For example, spreading a chemical film over swamps and

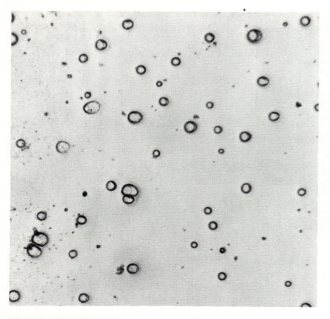

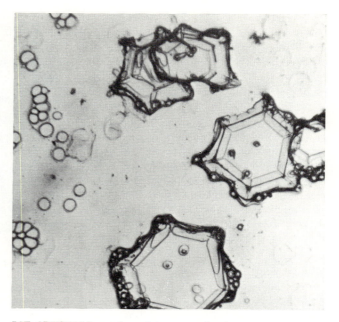

WATER DROPLETS that comprise a fog vary in size from about two microns to as much as 50 microns in diameter; their average size is 20 microns. The droplets in the photomicrograph are from a sample of supercooled fog prepared for a "cold seeding" study.

ICE CRYSTALS are formed in a sample of supercooled fog by seeding with propane. The crystals grow as water vapor is deposited on them. Reduced humidity makes the fog droplets evaporate, clearing the air and furnishing more water vapor for crystal growth.

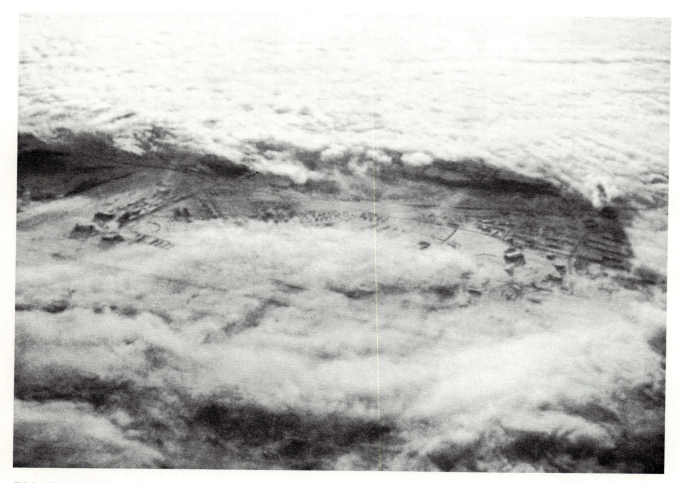

FOG CLEARANCE is achieved by dropping dry ice into a bank of supercooled fog overlying Elmendorf Air Force Base in Anchorage, Alaska. Seeding with dry ice initiated the growth of ice crystals, de- priving the air of the liquid water that had made it foggy. Only 5 percent of the fogs in those parts of the U.S. that lie within the Temperate Zone are supercooled and dispersable by cold-seeding.

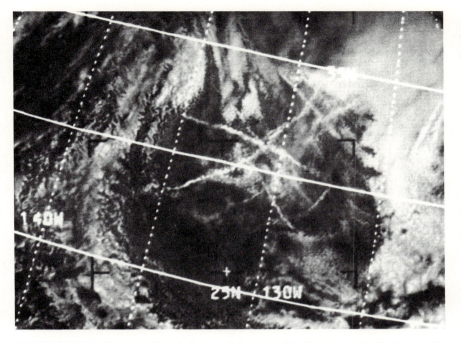

POLLUTION FOGS over the Pacific Ocean off the coast of California are visible in a weather satellite photograph as white trails in the area between 25 and 35 degrees north latitude and 125 and 135 degrees west longitude. Each trail is a narrow fogbank formed in response to the discharge of hygroscopic particles into the atmosphere from the funnels of ships at sea.

more careful attention to the selection of proper locations for activities that generate fog and for those that may be troubled by fog.

Factories giving rise to air pollution, for example, should be located at sites where the topography and prevailing winds favor effective dispersal of the pollutants. Plants that must dispose of heated water should not be built near densely populated or well-traveled areas. On the other hand, in the selection of sites for airports, highways, sports stadiums, golf courses and so forth consideration should be given to finding locations where the geography, topography, soil characteristics and other features would tend to minimize the formation and persistence of fogs. For example, ideally an airport should be situated (1) upwind of any nearby source of air pollution, (2) on a plateau that stands high enough to shed cold air into a valley but low enough to avoid hilltop immersion in clouds, (3) away from rivers, lakes or marshy ground and (4) in full, unobstructed exposure to the sun.

By the application of meteorological knowledge to planning and by the development of further methods for fog prevention and dispersal, we may eventually be able to deal with the growing fog and smog problem. The success of these efforts, however, will hinge on the achievement of effective control over the pollution of the air and waters.

lakes in the vicinity of sensitive areas such as airports and highways might considerably reduce evaporation and thus reduce the frequency and intensity of fogs in those areas. There are indications that in places where shallow radiation fogs are common the fogs could be prevented from spreading by planting vegetation thickly around the area of origin. It would be desirable to prohibit pollution-generating factories from operating at times when the air is calm or an inversion is present. And the fog menace could be diminished greatly by giving

BIBLIOGRAPHIES

1. The Growth of Snow Crystals

THE GROWTH OF ICE CRYSTALS FROM THE VAPOUR AND THE MELT. B. J. Mason in *Advances in Physics*, Vol. 7, No. 26, pages 235–253; April, 1958.

ICE-NUCLEATING PROPERTIES OF SOME NATURAL MINERAL DUSTS. B. J. Mason and J. Maybank in *Quarterly Journal of the Royal Meteorological Society*, Vol. 84, No. 361, pages 235–241; July, 1958.

THE INFLUENCE OF TEMPERATURE AND SUPERSATURATION ON ICE CRYSTALS GROWN FROM A VAPOR. J. Hallett and B. J. Mason in *Proceedings of the Royal Society*, Series A, Vol. 247, No. 1,251, pages 440–453; October 21, 1958.

THE PHYSICS OF CLOUDS. B. J. Mason. Oxford University Press, 1957.

THE SUPERCOOLING AND NUCLEATION OF WATER. B. J. Mason in *Advances in Physics*, Vol. 7, No. 26, pages 221–234; April, 1958.

2. The Aurora

THE AURORA. S. Akasofu, S. Chapman and A. B. Meinel in *Handbuch der Physik: Vol. XLIX*. Springer-Verlag, Inc., 1966, pages 1–158.

AURORAL PHENOMENA. B. J. O'Brien in *Science*, Vol. 148, No. 3669, pages 449–460; April 23, 1965.

DYNAMIC MORPHOLOGY OF AURORAS. Syun-Ichi Akasofu in *Space Science Reviews*, Vol. 4, No. 4, pages 498–540; June, 1965.

POLAR AURORAS. V. I. Krasovskij in *Space Science Reviews*, Vol. 3, No. 2, pages 232–274; 1964.

3. The Shape of Raindrops

THEORETICAL CLOUD PHYSICS STUDIES. Final Report on Contract Nonr757, Project No. NRO82093. Office of Naval Research, January 31, 1953.

4. River Meanders

FLUVIAL PROCESSES IN GEOMORPHOLOGY. Luna B. Leopold, M. Gordon Wolman and John P. Miller. W. H. Freeman and Company, 1964

RIVER MEANDERS. Luna B. Leopold and M. Gordon Wolman in *Bulletin of the Geological Society of America*, Vol. 71, No. 6, pages 769–794; June, 1960.

RIVERS. Luna B. Leopold in *American Scientist*, Vol. 50, No. 4, pages 511–537; December, 1962.

5. Thunder

LIGHTNING CHANNEL RECONSTRUCTION FROM THUNDER MEASUREMENTS. A. A. Few in *Journal of Geophysical Research*, Vol. 75, No. 36, pages 7517–7523; December 20, 1970.

THUNDER SIGNATURES. A. A. Few in *E ⊕ S Transactions of the American Geophysical Union*, Vol. 55, No. 5, pages 508–514; May, 1974.

HORIZONTAL LIGHTNING. Thomas L. Teer and A. A. Few in *Journal of Geophysical Research*, Vol. 79, No. 24, pages 3436–3441; August 20, 1974.

6. Why the Sea Is Salt

THE OCEANS: THEIR PHYSICS, CHEMISTRY, AND GENERAL BIOLOGY. H. U. Sverdrup, Martin W. Johnson and Richard H. Fleming. Prentice-Hall, Inc., 1961.

THE COMPOSITION OF SEA-WATER, SECTION I: CHEMISTRY in *The Sea: Ideas and Observations on Progress in the Study of the Seas, Vol. II*, edited by M. N. Hill. Interscience Publishers, 1963.

CHEMICAL OCEANOGRAPHY. Edited by J. P. Riley and G. Skirrow. Academic Press, 1965.

THE OCEAN AS A CHEMICAL SYSTEM. Lars Gunnar Sillén in *Science*, Vol. 156, No. 3779, pages 1189–1197; June 2, 1967.

MARINE CHEMISTRY: THE STRUCTURE OF WATER AND THE CHEMISTRY OF THE HYDROSPHERE. R. A. Horne. Wiley-Interscience, 1969.

7. Ocean Waves

BREAKERS AND SURF: PRINCIPLES IN FORECASTING. Hydrographic Office Publication No. 234, 1944.

THE OCEANS: THEIR PHYSICS, CHEMISTRY AND GENERAL BIOLOGY. H. U. Sverdrup, Martin W. Johnson and Richard H. Fleming. Prentice-Hall, Inc., 1942.

PRACTICAL METHODS FOR OBSERVING AND FORECASTING OCEAN WAVES BY MEANS OF WAVE SPECTRA AND STATISTICS. Willard J. Pierson, Jr., Gerhard Neuman and Richard W. James. Hydrographic Office Publication No. 603, 1958.

8. Fog

FOG. Joseph J. George in *Compendium of Meteorology,* edited by Thomas F. Malone. American Meteorological Society, 1951.

WEATHER ANALYSIS AND FORECASTING, VOL. II: WEATHER AND WEATHER SYSTEMS. Sverre Petterssen. McGraw-Hill Book Company, Inc., 1956.

CLOUD PHYSICS AND CLOUD SEEDING. Louis J. Battan. Anchor Books, Doubleday & Company, Inc., 1962.

THE UNCLEAN SKY: A METEOROLOGIST LOOKS AT AIR POLLUTION. Louis J. Battan. Anchor Books, Doubleday & Company, Inc., 1966.

AIR POLLUTION. 1968. R. S. Scorer. Pergamon Press, 1968.

INDEX